Emulsions and Oil Treating Equipment

Emulsions and Oil Treating Equipment

Selection, Sizing and Troubleshooting

Maurice Stewart
Ken Arnold

ELSEVIER

AMSTERDAM • BOSTON • HEIDELBERG • LONDON • NEW YORK • OXFORD
PARIS • SAN DIEGO • SAN FRANCISCO • SINGAPORE • SYDNEY • TOKYO

Gulf Professional Publishing is an imprint of Elsevier

Gulf Professional Publishing is an imprint of Elsevier
30 Corporate Drive, Suite 400, Burlington, MA 01803, USA
Linacre House, Jordan Hill, Oxford OX2 8DP, UK

Library of Congress Cataloging-in-Publication Data
Application submitted

British Library Cataloguing-in-Publication Data
A catalogue record for this book is available from the British Library.

ISBN: 978-0-7506-8970-0

For information on all Gulf Professional Publishing publications
visit our Web site at www.elsevierdirect.com

Printed and bound in the United Kingdom
Transferred to Digital Printing, 2010

Contents

A Note from the Authors

Gulf Equipment Guides series serves as a quick reference for the design, selection, specification, installation, operation, testing, and trouble-shooting of surface production equipment. The *Gulf Equipment Guides* series consists of multiple volumes, each of which covers a specific area in surface production equipment. These guides cover essentially the same topics included in the "Surface Production Operations" series but omit the proofs of equations, example problems and solutions which belong more properly in a handbook. This book contains fewer pages and is therefore more focused. The reader is referred to the corresponding volume of the "Surface Production Operations" series for further details and additional information such as derivations of some of the equations, example problems and solutions and suggested test questions.

About the Book

Emulsions and Oil Treating Equipment: Selection, Sizing and Troubleshooting is the second volume in the *Gulf Equipment Guides* series. Each guide serves as a quick reference resource. The series is intended to provide the most comprehensive coverage you'll find today dealing with surface production operation in its various stages, from initial entry into the flowline through gas–liquid and liquid–liquid separation; emulsions, oil and water treating; water injection; hydrate prediction and prevention; gas dehydration; and gas conditioning and processing equipment to the exiting pipeline.

Featured in this volume are such important topics as emulsions, oil treating, desalting, water treating, water injection systems, and any other related topics. This volume as well as all volumes in the *Gulf Equipment Guides* series, serve the practicing engineer and senior field personnel by providing organized design procedures; details on suitable equipment for application selection; and charts, tables, and nomographs in readily useable form. Facility engineers, process engineers, designers, operations engineers, and senior production operators will develop a "feel" for the important parameters in designing, selecting, specifying, and trouble-shooting surface production facilities. Readers will understand the uncertainties and assumptions inherent in designing and operating the equipment in these systems and the limitations, advantages, and disadvantages associated with their use.

CHAPTER 1

Crude Oil Treating Systems

1.1 Introduction

Conditioning of oil-field crude oils for pipeline quality is complicated by water produced with the oil. Separating water out of produced oil is performed by various schemes with various degrees of success. The problem of removing emulsified water has grown more widespread and oftentimes more difficult as production schemes lift more water with oil from water-drive formations, water-flooded zones, and wells stimulated by thermal and chemical recovery techniques. This chapter describes oil-field emulsions and their characteristics, treating oil-field emulsions so as to obtain pipeline quality oil, and equipment used in conditioning oil-field emulsions.

1.2 Equipment Description

1.2.1 Free-Water Knockouts

Most well streams contain water droplets of varying size. If they collect together and settle to the bottom of a sample within 3–10 min, they are called "free water." This is an arbitrary definition, but it is generally used in designing equipment to remove water that will settle out rapidly. A free-water knockout (FWKO) is a pressure vessel used to remove free water from crude oil streams (Figure 1.1). They are located in the production flow path where turbulence has been minimized. Restrictions such as orifices, chokes, throttling globe valves, and fittings create turbulence in the liquids that aggravate emulsions. Free water, at wellhead conditions, frequently will settle out readily to the bottom of an expansion chamber.

Sizing and pressure ratings for these vessels are discussed in the "Gas–Liquid and Liquid–Liquid Separation" volume, this series. Factors affecting design include retention time, flow rate (throughput), temperature, oil gravity (as it influences viscosity), water drop size

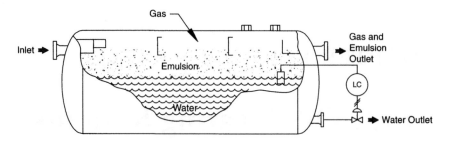

FIGURE 1.1. Cutaway of a free-water knockout.

distribution, and emulsion characteristics. Abnormal volumes of gas in the inlet stream may require proportionately larger vessels as these gas volumes affect the throughput rate. A simple "field check" to determine retention time is to observe a fresh sample of the wellhead crude and the time required for free water to segregate.

In installations where gas volumes vary, a two-phase separator is usually installed upstream of the FWKO. The two-phase separator removes most of the gas and reduces turbulence in the FWKO vessel. The FWKO usually operates at 50 psig (345 kPa) or less due to the vessel's location in the process flow stream. Internals should be coated or protected from corrosion since they will be in constant contact with salt water.

1.2.2 Gunbarrel Tanks with Internal and External Gas Boots

The gunbarrel tank, sometimes called a wash tank, is the oldest equipment used for multi-well onshore oil treating in a conventional gathering station or tank battery. Gunbarrel tanks are very common in heavy crude applications such as in Sumatra and East Kalimantan, Indonesia, and in Bakersfield, California.

The gunbarrel tank is a vertical flow treater in an atmospheric tank. Figure 1.2 shows a "gunbarrel" tank with an internal gas boot. Typically, gunbarrels have an internal gas separating chamber or "gas boot" extending 6–12 ft (2–4 m) above the top of the tank, where gas is separated and vented, and a down-comer extending 2–5 ft (0.6–1.5 m) from the bottom of the tank. A variation of the above gunbarrel configuration is a wash tank with an "external" gas boot. This configuration is preferred on larger tanks, generally in the 60,000-barrel range, where attaching an internal gas boot is structurally difficult. In either case, the gunbarrel tank is nothing more than a large atmospheric settling tank that is higher than downstream oil shipping and water clarifier tanks. The elevation difference allows gravity flow into the downstream vessels.

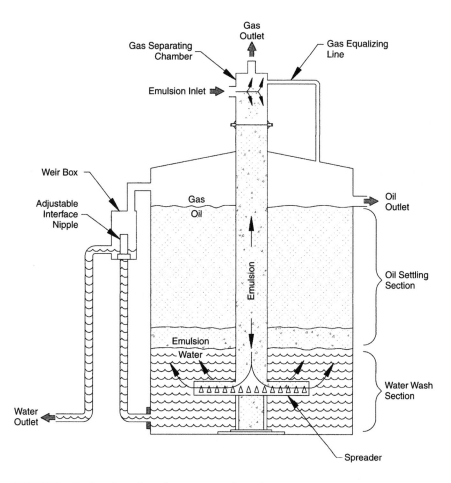

FIGURE 1.2. Gunbarrel with an internal gas boot.

Because gunbarrels tend to be of larger diameter than vertical heater-treaters, many have elaborate spreader systems that attempt to create uniform (i.e., plug) upward flow of the emulsion to take maximum advantage of the entire cross section. Spreader design is important to minimize the amount of short-circuiting in larger tanks.

The emulsion, flowing from an upstream separator and possibly a heater, enters the top of the gas separation section of the gas boot. The gravity separation section removes flash gas and gas liberated as a result of heating the emulsion. The emulsion flows down the down-comer to a spreader, which is positioned below the oil–water interface. Exiting at the bottom of the down-comer, the emulsion rises to the top of the surrounding layer of water. The water level is controlled by a water leg or automatic level control. The emulsion passage through the water helps collect the entrained water and converts the emulsion into distinct oil and water layers. Oil accumulates at the top and flows out through the spillover line into the oil settling tank. Water flows from the bottom of the tank, up through the water leg, and into a surge or clarifier tank. The height of the water leg regulates the amount of water retained in the vessel. The settling time in the vessel for the total fluid stream is usually 12–24 h. Most gunbarrels are unheated, though it is possible to provide heat by heating the incoming stream external to the tank, installing heating coils in the tank, or circulating the water to an external or 'jug' heater in a closed loop. It is preferable to heat the inlet so that more gas is liberated in the boot, although this means that fuel will be used in heating any free water in the inlet.

The height of the external water leg controls the oil–water interface inside the vessel and automatically allows clean oil and produced water to exit the vessel. Example 1.1 illustrates this design consideration.

1.2.3 Determination of External Water Leg Height
Given:

Oil gravity at 60 °F	36 °API
Water specific gravity	1.05
Height of oil outlet	23 ft
Height of interface level	10 ft (for this example)
Height of water outlet	1 ft
Figure 1.3	Gunbarrel schematic

Solution:
Determine the oil specific gravity.

$$\text{Oil specific gravity} = \frac{141.5}{131.5 + {}^{\circ}\text{API}} = \frac{141.5}{131.5 + 36} = 0.845$$

1. Determine the oil gradient.

Since the charge in the pressure with depth for fresh water is 0.433 psi/ft of depth, the change in pressure with depth of fluid whose specific gravity is SG would be 0.433 (SG); thus, the oil gradient is

$$\text{Oil gradient} = (0.433)(0.845) = 0.366 \text{ psi/ft.}$$

2. Determine the water gradient.

$$\text{Water gradient} = (0.433)(1.05) = 0.455 \text{ psi/ft.}$$

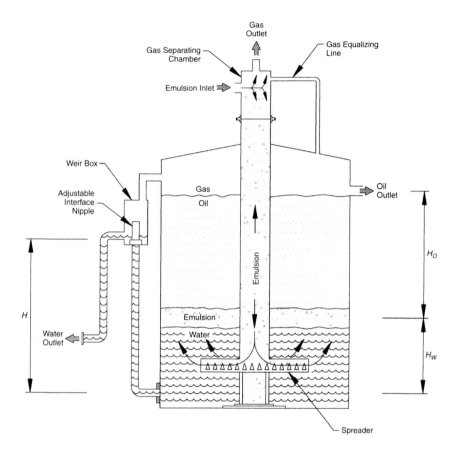

FIGURE 1.3. Determination of external water leg height, H.

3. Calculate the height of the oil and the height of the water in the tank.

H_o = height of oil outlet – height of interface level = $23 - 10 = 13$ ft

H_w = height of interface level – height of water outlet = $10 - 1 = 9$ ft.

4. Perform a pressure balance.

$$\left(\begin{array}{c} \text{Hydrostatic pressure} \\ \text{inside tank} \end{array}\right) = \left(\begin{array}{c} \text{Hydrostatic pressure} \\ \text{in the water leg} \end{array}\right),$$

$$(13)(0.366) + (9)(0.455) = (H)(0.455),$$

$$H = \frac{(13)(0.366) + (9)(0.455)}{0.455} = 19.5 \text{ ft}.$$

The design details for the spreader, water leg, and gas separation section vary for different manufacturers. These details do not significantly affect the sizing of the tank, provided the spreader minimizes short-circuiting. No matter how careful the design of the spreaders, large wash tanks are very susceptible to short-circuiting. This is due to temperature and density differences between the inlet emulsion and the fluid in the tank, solids deposition, and corrosion of the spreaders.

Standard tank dimensions are listed in API Specification 12F (Shop Welded Tanks), API Specification 12D (Field Welded Tanks), and API Specification 12B (Bolted Tanks). These dimensions are shown in Tables 1.1, 1.2, and 1.3, respectively.

Gunbarrels are simple to operate and, despite their large size, are relatively inexpensive. However, they have a large footprint, which is why they are not used on offshore platforms. Gunbarrels hold a large

TABLE 1.1
Shop welded tanks (API specification 12 F)

(a) Field Units				
Nominal Capacity (bbl)	Approximate Working Capacity (bbl)	Outside Diameter (ft–in.)	Height (ft–in.)	Height of Overflow Connection (ft–in.)
90	72	7–11	10	9–6
100	79	9–6	8	7–6
150	129	9–6	12	11–6
200	166	12–0	10	9–6
210	200	10–0	15	14–6
250	224	11–0	15	14–6
300	266	12–0	15	14–6
500	479	15–6	16	15–6

(b) SI Units				
Nominal Capacity (bbl)	Approximate Working Capacity (m^3)	Outside Diameter (m)	Height (m)	Height of Overflow Connection (m)
90	11.4	2.41	3.05	2.90
100	12.6	2.90	2.44	2.29
150	20.5	2.90	3.66	3.51
200	26.4	3.66	3.05	2.90
210	31.8	3.05	4.57	4.42
250	35.6	3.35	4.57	4.42
300	42.3	3.66	4.57	4.42
500	76.2	4.72	4.88	4.72

TABLE 1.2
Field welded tanks (API specification 12D)

(a) Field Units

Nominal Capacity (bbl)	Design Pressure (oz/in.²)		Approximate Working Capacity (bbl)	Nominal Outside Diameter (ft–in.)	Nominal Height (ft–in.)	Height of Overflow Line Connection (ft–in.)
	Pressure	Vacuum				
H-500	8	1/2	479	15–6	16–0	15–6
750	8	1/2	746	15–6	24–0	23–6
L-500	6	1/2	407	21–6	8–0	7–6
H-1000	6	1/2	923	21–6	16–0	15–6
1500	6	1/2	1438	21–6	24–0	23–6
L-1000	4	1/2	784	29–9	8–0	7–6
2000	4	1/2	1774	29–9	16–0	15–6
3000	4	1/2	2764	29–9	24–0	23–6
5000	3	1/2	4916	38–8	24–0	23–6
10,000	3	1/2	9938	55–0	24–0	23–6

(b) SI Units

Nominal Capacity (bbl)	Design Pressure (kPa)		Approximate Working Capacity (m³)	Nominal Outside Diameter (m)	Nominal Height (m)	Height of Overflow Line Connection (m)
	Pressure	Vacuum				
H-500	3.4	0.2	76.2	4.72	4.88	4.72
750	3.4	0.2	118.6	4.72	7.32	7.16
L-500	2.6	0.2	64.7	6.55	2.44	2.29
H-1000	2.6	0.2	146.8	6.55	4.88	4.72
1500	2.6	0.2	228.6	6.55	7.32	7.16
L-1000	1.7	0.2	124.6	9.07	2.44	2.29
2000	1.7	0.2	282.0	9.07	4.88	4.72
3000	1.7	0.2	439.4	9.07	7.32	7.16
5000	1.3	0.2	781.6	11.79	7.32	7.16
10,000	1.3	0.2	1580.0	16.76	7.32	7.16

volume of fluids, which is a disadvantage should a problem develop. When the treating problem is detected in the oil outlet, a large volume of bad oil has already collected in the tank. This oil may have to be treated again, which may require large slop tanks, recycle pumps, etc. It may be beneficial to reprocess this bad oil in a separate treating facility so as to avoid further contamination of the primary treating facility.

Gunbarrels are most often used in older, low-flow-rate, onshore facilities. In recent times, vertical heater-treaters have become so inexpensive that they have replaced gunbarrels in single-well

TABLE 1.3
Bolted tanks (API specification 12B)

		(a) Field Units		
Nominal Capacity (42-gal bbl)	*Number of Rings*	*Inside Diameter[a] (ft–in.)*	*Height of Shell[b] (ft–in.)*	*Calculated Capacity[c] (42-gal bbl)*
100	1	$9\text{–}2\frac{3}{4}$	$8\frac{1}{2}$	96
200	2	$9\text{–}2\frac{3}{4}$	16–1	192
300	3	8–2	$24\text{–}1\frac{1}{2}$	287
250	1	$15\text{–}4\frac{5}{8}$	$8\frac{1}{2}$	266
High 500	2	$15\text{–}4\frac{5}{8}$	16–1	533
750	3	$15\text{–}4\frac{5}{8}$	$24\text{–}1\frac{1}{2}$	799
Low 500	1	$21\text{–}6\frac{1}{2}$	$8\frac{1}{2}$	522
High 1000	2	$21\text{–}6\frac{1}{2}$	16–1	1044
1500	3	21–6	$24\text{–}1\frac{1}{2}$	1566
Low 1000	1	$29\text{–}8\frac{5}{8}$	$8\frac{1}{2}$	944
2000	2	$29\text{–}8\frac{5}{8}$	16–1	1987
3000	3	$29\text{–}8\frac{5}{8}$	$24\text{–}1\frac{1}{2}$	2981
5000	3	$38\text{–}7\frac{5}{8}$	$24\text{–}1\frac{1}{2}$	5037
10,000	3	$54\text{–}11\frac{3}{4}$	24–2	10,218
		(b) SI Units		
Nominal Capacity (42-gal bbl)	*Number of Rings*	*Inside Diameter[d] (m)*	*Height of Shell[b] (m)*	*Calculated Capacity[c] (m³)*
100	1	2.81	2.45	15.3
200	2	2.81	4.90	30.5
300	3	2.81	7.35	45.6
250	1	4.69	2.45	42.3
High 500	2	4.69	4.90	84.7
750	3	4.69	7.35	127.0
Low 500	1	6.57	2.45	83.0
High 1000	2	6.57	4.90	166.0
1500	3	6.57	7.35	249.0
Low 1000	1	9.06	2.45	158.0
2000	2	9.06	4.90	315.9
3000	3	9.06	7.30	473.9
5000	3	11.78	7.30	800.8
10,000	3	16.76	7.37	1624.5

[a]The inside diameter is an approximate dimension. The values shown are 2 in. less than the bottom bolt-circle diameters.
[b]Shell heights shown do not include the thickness of the gasket.
[c]The calculated capacity is based on the inside diameter and height of shell.
[d]The inside diameter is an approximate dimension. The values shown are less than the bottom bolt-circle diameters.

installations. On larger installations onshore in warm weather areas, gunbarrels are still commonly used. In areas that have a winter season they tend to become too expensive to keep the large volume of oil at a high enough temperature to combat potential pour-point problems.

1.2.4 Horizontal Flow Treaters

Horizontal flow treaters are not common. Figure 1.4 illustrates one design, which consists of a cylindrical treating tank incorporating internal baffles. The internal baffles establish a horizontal flow pattern in the cylindrical tank, which is more efficient for gravity separation than vertical flow and is less subject to short-circuiting.

The oil, emulsion, and water enter the vessel and must follow the long flow path between the baffles. Separation takes place in the straight flow areas between the baffles. Turbulence coupled with high flow velocities prevents separation at the corners, where the flow reverses direction. Tracer studies indicate that approximately two thirds of the plan area of the tank is effective in oil–water separation.

In addition to gravity separation, the emulsion must be collected and held in the treater for a certain retention time so that the emulsion will break. In horizontal flow treaters, the emulsion collects

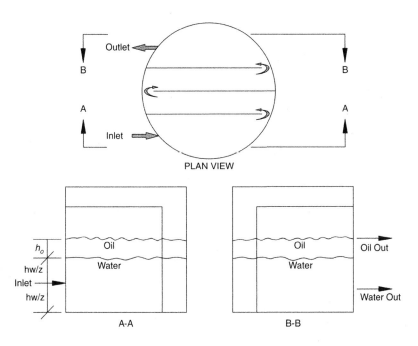

FIGURE 1.4. Plan view of a cylindrical treating tank incorporating internal baffles that establish horizontal flow.

between the oil and water; however, the horizontal flow pattern tends to sweep the emulsion toward the outlets. The emulsion layer may grow much thicker at the outlet end of the treater than at the inlet end. Accordingly, it is much easier for the emulsion to be carried out of the vessel with the oil.

1.2.5 Heaters

Heaters are vessels used to raise the temperature of the liquid before it enters a gunbarrel, wash tank, or horizontal flow treater. They are used to treat crude oil emulsions. The two types of heaters commonly used in upstream operations are indirect fired heaters and direct fired heaters. Both types have a shell and a fire tube. Indirect heaters have a third element, which is the process flow coil. Heaters have standard accessories such as burners, regulators, relief valves, thermometers, temperature controllers, etc.

Indirect Fired Heaters

Figure 1.5 shows a typical indirect fired heater. Oil flows through tubes that are immersed in water, which in turn is heated by a fire tube. The heat may be supplied by a heating fluid medium, steam, or electric immersed heaters. Indirect heaters maintain a constant temperature over a long period of time and are safer than the direct heater. Hot spots are not as likely to occur if the calcium content of the heating water is controlled. The primary disadvantage is that these heaters require several hours to reach the desired temperature after they have been out of service.

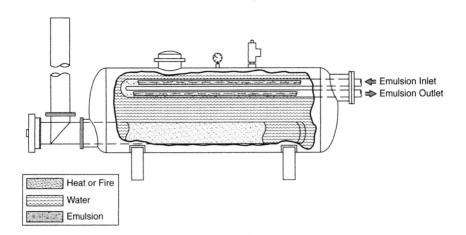

FIGURE 1.5. Cutaway of a horizontal indirect fired heater.

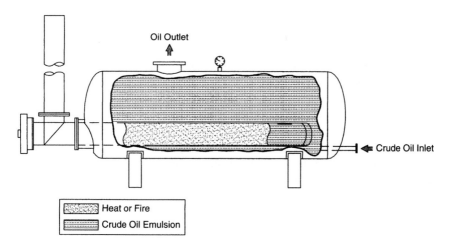

FIGURE 1.6. Cutaway of a horizontal direct fired heater.

Direct Fired Heaters

Figure 1.6 shows a typical direct fired heater. Oil flows through an inlet distributor and is heated directly by a fire box. The heat may be supplied by a heating fluid medium, steam, or an electric immersed heater. Direct heaters are quick to reach the desired temperature, are efficient (75–90%), and offer a reasonable initial cost. Direct fired heaters are typically used where fuel gas is available and high volume oil treating is required. On the other hand, they are hazardous and require special safety equipment. Scale may form on the oil side of the fire tube, which prevents the transfer of heat from the fire box to the oil emulsion. Heat collects in the steel walls under the scale, which causes the metal to soften and buckle. The metal eventually ruptures and allows oil to flow into the fire box, which results in a fire. The resultant blaze, if not extinguished, will be fed by the incoming oil stream.

1.2.6 Waste Heat Recovery

A waste heat recovery heater captures waste heat from the exhaust stacks of compressors, turbines, generators, and large engines. Heat exchangers are used to transfer this heat to a heating fluid medium, which in turn is used to heat the crude oil emulsion.

1.2.7 Heater-Treaters

Heater-treaters are an improvement over the gunbarrel and heater system. Many designs are offered to handle various conditions such as viscosity, oil gravity, high and low flow rates, corrosion, and cold weather. When compared to gunbarrels, heater-treaters are less expensive initially, offer lower installation costs, provide greater heat efficiency, provide greater flexibility, and experience greater overall efficiency. On the other hand, they are more complicated, provide less storage space for basic sediment, and are more sensitive to chemicals. Since heater-treaters are smaller than other treating vessels, their retention times are minimal (10–30 min) when compared to gunbarrels and horizontal flow treaters.

Internal corrosion of the down-comer pipe is a common problem. Build-up of sediment on the walls or bottom of the treater can cause the interface levels to rise and liquid to carry over and/or oil to exit the treater with salt water. Bi-annual inspections should be performed to include internal inspection for corrosion, sediment build-up, and scale build-up.

1.2.8 Vertical Heater-Treaters

The most commonly used single-well treater is the vertical heater-treater, which is shown in Figure 1.7. The vertical heater-treater consists of four major sections: gas separation, FWKO, heating and water-wash, and coalescing-settling sections. Incoming fluid enters the top of the treater into a gas separation section, where gas separates from the liquid and leaves through the gas line. Care must be exercised to size this section so that it has adequate dimensions to separate the gas from the inlet flow. If the treater is located downstream of a separator, the gas separation section can be very small. The gas separation section should have an inlet diverter and a mist extractor.

The liquids flow through a down-comer to the base of the treater, which serves as a FWKO section. If the treater is located downstream of a FWKO or a three-phase separator, the bottom section can be very small. If the total wellstream is to be treated, this section should be sized for 3–5 min retention time to allow the free water to settle out. This will minimize the amount of fuel gas needed to heat the liquid stream rising through the heating section. The end of the down-comer should be slightly below the oil–water interface so as to 'water-wash' the oil being treated. This will assist in the coalescence of water droplets in the oil.

The oil and emulsion rise through the heating and water-wash section, where the fluid is heated (Figure 1.8). As shown in Figure 1.9, a fire tube is commonly used to heat the emulsion in the heating and

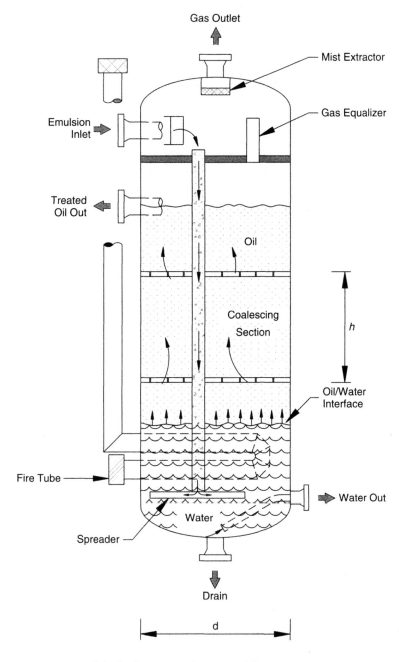

FIGURE 1.7. Simplified schematic of a vertical heater-treater.

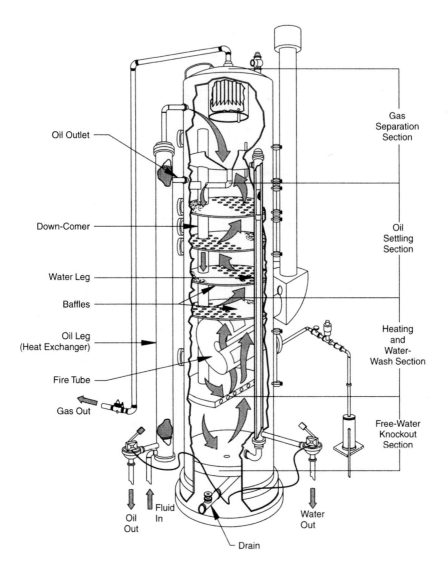

FIGURE 1.8. Three-dimensional view illustrating oil and emulsion rising through the heating and water-wash.

water-wash section. After the oil and emulsion are heated, the heated oil and emulsion enter the coalescing section, where sufficient retention time is provided to allow the small water droplets in the oil continuous phase to coalesce and settle to the bottom. As shown in Figure 1.10, baffles are sometimes installed in the coalescing section to treat difficult emulsions. The baffles cause the oil and emulsion

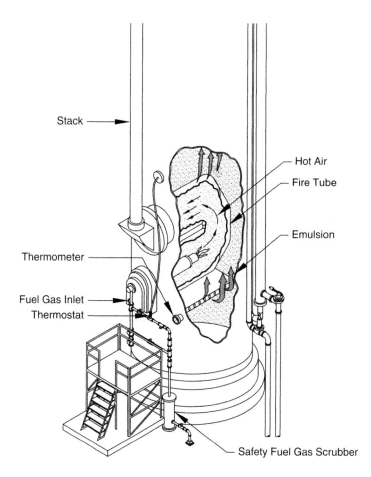

Stack

Hot Air

Fire Tube

Emulsion

Thermometer

Fuel Gas Inlet

Thermostat

Safety Fuel Gas Scrubber

FIGURE 1.9. Cutaway showing a typical fire-tube that heats the emulsion in the heating and water-wash section.

to follow a back-and-forth path up through the treater. Heating causes more gas to separate from the oil than is captured in the condensing head. Treated oil flows out the oil outlet, at the top of the coalescing section, and through the oil leg heat exchanger, where a valve controls the flow. Heated clean oil preheats incoming cooler emulsion in the oil leg heat exchanger (Figure 1.11). Separated water flows out through the water leg, where a control valve controls the flow to the water treating system (Figure 1.12).

As shown in Figure 1.13, any gas flashed from the oil due to heating, is captured on the condensing head. Any gas that did not condense flows through an equalizing line to the gas separation section.

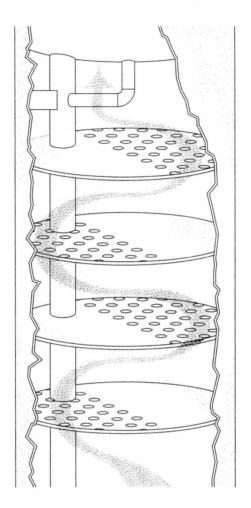

FIGURE 1.10. Baffles, installed in the coalescing section, cause the emulsion to follow a back-and-forth path up through the oil settling section.

As shown in Figure 1.14, a vane-type mist extractor removes the liquid mist before the gas leaves the treater. The gas liberated when crude oil is heated may create a problem in the treater if it is not adequately designed. In vertical heater-treaters the gas rises through the coalescing section. If a great deal of gas is liberated, it can create enough turbulence and disturbance to inhibit coalescence. Equally important is the fact that small gas bubbles have an attraction for surface-active material and hence water droplets. Thus, they tend to keep

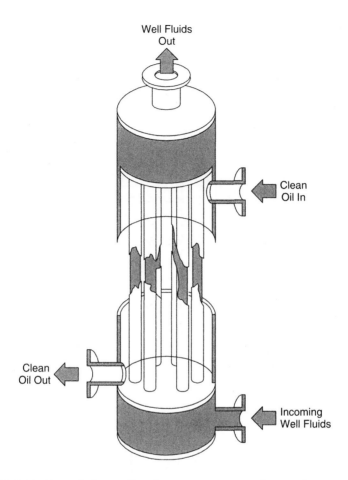

FIGURE 1.11. Heated clean oil preheats incoming cooler emulsion in the oil leg heat exchanger.

the water droplets from settling out and may even cause them to carry over to the oil outlet.

The oil level is maintained by pneumatic or lever-operated dump valves. The oil–water interface is controlled by an interface level controller or an adjustable external water leg.

Standard vertical heater-treaters are available in 20- and 27-ft (6.1 and 8.2 m) heights. These heights provide sufficient static liquid head so as to prevent vaporization of the oil. The detailed design of the treater, including the design of internals (many features of which are patented), should be the responsibility of the equipment supplier.

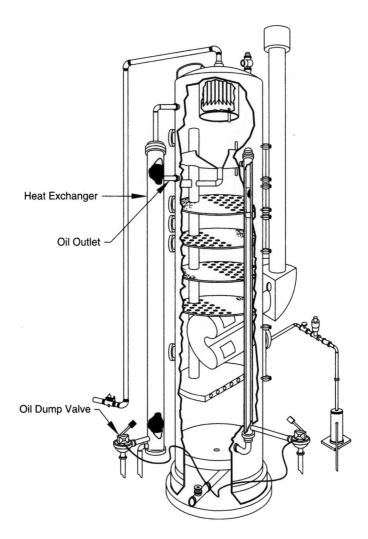

FIGURE 1.12. Cutaway illustrating oil and water legs.

1.2.9 Coalescing Media

It is possible to use coalescing media to promote coalescence of the water droplets. These media provide large surface areas upon which water droplets can collect. In the past the most commonly used coalescing media was wood shavings or 'excelsior,' which is also referred to as a 'hay section.' The wood excelsior was tightly packed to create an obstruction to the flow of the small water droplets and promote random collision of these droplets for coalescence. When the droplets were large enough, they fell out of the flow stream by gravity. Figure 1.15 shows a vertical heater-treater utilizing an excelsior.

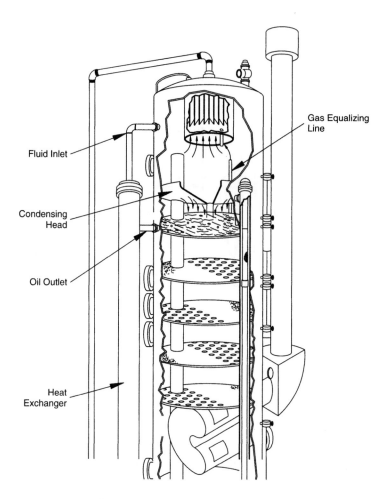

FIGURE 1.13. Gas, flashed from the oil during heating, is captured on the condensing head.

The use of an "excelsior" allowed lower treating temperatures. However, these media had a tendency to clog with time and were difficult to remove. Therefore, they are no longer used.

1.2.10 Horizontal Heater-Treaters

For most multi-well flow streams, horizontal heater-treaters are normally required. Figure 1.16 shows a simplified schematic of a typical horizontal heater-treater. Design details vary from manufacturer to manufacturer, but the principles are the same. The horizontal heater-treater consists of three major sections: front (heating and water-wash), oil surge chamber, and coalescing sections.

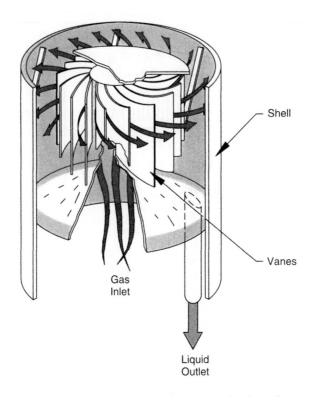

FIGURE 1.14. Vane-type mist extractor removes the liquid mist before the gas leaves the treater.

Incoming fluids enter the front (heating and water-wash) section through the fluid inlet and down over the deflector hood (Figure 1.17) where gas is flashed and removed. Heavier materials (water and solids) flow to the bottom while lighter materials (gas and oil) flow to the top. Free gas breaks out and passes through the gas equalizer loop to the gas outlet. As shown in Figure 1.18, the oil, emulsion, and free water pass around the deflector hood to the spreader located slightly below the oil–water interface, where the liquid is "water-washed" and the free water is separated. For low gas–oil-ratio crudes, blanket gas may be required to maintain gas pressure. The oil and emulsion are heated as they rise past the fire tubes and are skimmed into the oil surge chamber.

As free water separates from the incoming fluids in the front section, the water level rises. If the water is not removed, it will continue to rise until it displaces all emulsion and begins to spill over the weir into the surge section. On the other hand, if the water level becomes too low, the front section will not be able to water-wash the incoming oil and emulsion, which would reduce the efficiency of the treater. Therefore, it is important to accurately control the oil–water interface

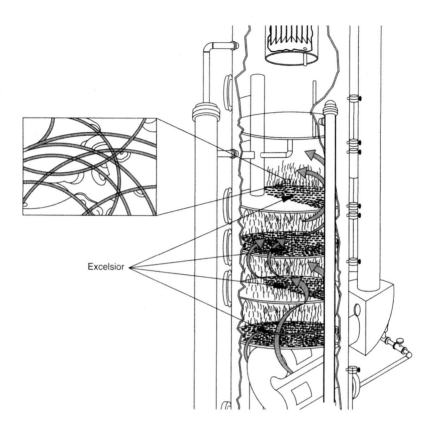

Excelsior

FIGURE 1.15. Vertical heater-treater fitted with excelsior, between the baffles, which aids in coalescence of water droplets.

in the front section. The oil–water interface is controlled by either an interface level controller, which operates a dump valve for the free water (Figure 1.19), or a resistance probe. If the water outlet valve sticks open, all the water and oil run out, exposing the fire tube or heat source.

As shown in Figure 1.20, a level safety low shutdown sensor is required in the upper portion of the front (heating and water-wash) section. This sensor assures liquid is always above the fire tube. If the water dump valve malfunctions or fails open, the liquid surrounding the fire tube will drop, thus not absorbing the heat generated from the fire tube and possibly damaging the fire tube by overheating. Thus, if the level above the fire tube drops, the level safety low shutdown sensor sends a signal that closes the fuel valve feeding the fire tube. It is also important to control the temperature of the fluid in the front (heating and water-wash) section. Therefore, a temperature controller, controlling the fuel to the burner or heat source, is required in the upper part of the heating–water-wash section (Figure 1.21).

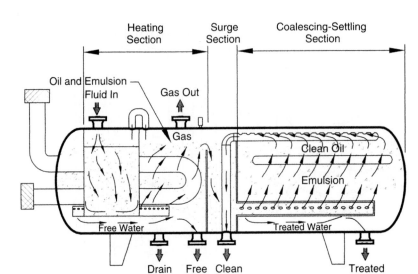

FIGURE 1.16. Simplified schematic of a horizontal heater-treater.

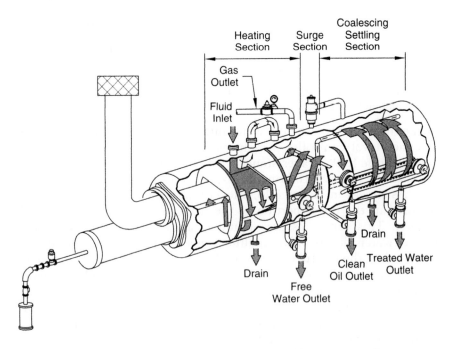

FIGURE 1.17. Three-dimensional view of a horizontal heater-treater flow pattern.

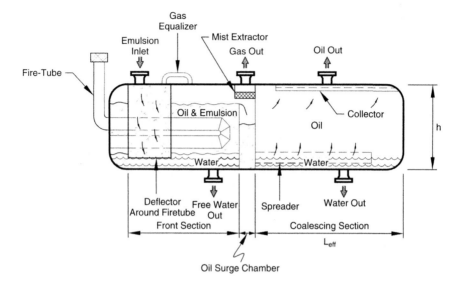

FIGURE 1.18. Schematic of horizontal heater-treater showing the oil, emulsion, and free water passing around the deflector hood to the spreader located slightly below the oil–water interface where the liquid is "water-washed" and the free water separated.

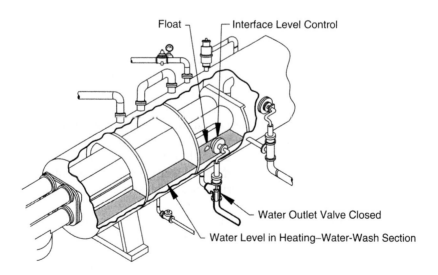

FIGURE 1.19. Oil–water interface in the heating and water-wash section is controlled by an interface level controller.

Level Safety Low Fuel
Shutdown Sensor

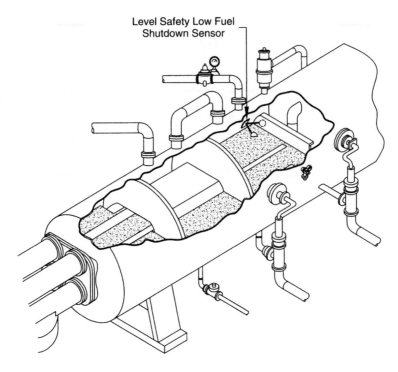

FIGURE 1.20. Level safety low sensor, located at the top of the heating–water-wash section, shuts off the fuel to the heat source (fire-tube) on low liquid level.

A level controller, in the oil surge section (Figure 1.22), operates the dump valve on the clean oil outlet line. This dump valve regulates the flow of oil out the top of the vessel, which maintains a liquid packed condition. When the clean oil outlet valve is open, the pressure of the gas in the surge chamber forces the emulsion to flow through the spreader and push the clean oil outlet through the clean oil collector (Figure 1.23). When the clean oil outlet valve closes, the flow of emulsion to the coalescing-settling section stops, and gas is prevented from entering the coalescing-settling section (Figure 1.24).

The oil and emulsion flow through a spreader into the back or coalescing section of the vessel, which is fluid packed. The spreader distributes the flow evenly throughout the length of this section. Because it is lighter than the emulsion and water, treated oil rises to the clean oil collector, where it is collected and passes the treater through the clean oil outlet. The collector is sized to maintain uniform vertical flow of the oil. Coalescing water droplets fall countercurrent to the rising oil continuous phase.

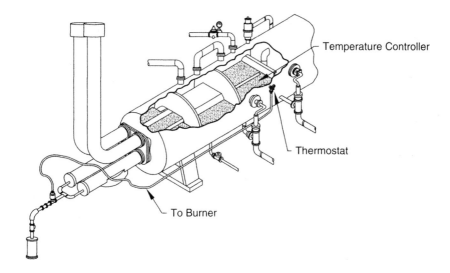

FIGURE 1.21. Temperature controller, located in the upper part of the heating–water-wash section, controls the fuel to the burner or heat source.

The front (heating and water-wash) section must be sized to handle settling of the free water and heating of the oil. The coalescing section must be sized to provide adequate retention time for coalescing to occur and to allow the coalescing water droplets to settle downward countercurrent to the upward flow of the oil.

Most horizontal heater-treaters built today do not use fire-tubes. Heat is added to the emulsion in a heat exchanger before the emulsion enters the treater. In these cases the inlet section of the treater can be fairly short because its main purpose is to degas the emulsion before it flows to the coalescing section.

Some heater-treaters are designed with only the coalescing section. In these cases the inlet is pumped through a heat exchanger to a treater that operates at a high enough pressure to keep the oil above its bubble-point. Thus, the gas will not evolve in the coalescing section of the treater.

1.2.11 Electrostatic Heater-Treaters

Some horizontal heater-treaters add an electrostatic grid in the coalescing section. Figure 1.25 illustrates a simplified schematic of a typical horizontal electrostatic treater. The flow path in an electrostatic heater-treater is basically the same as in a horizontal heater-treater, except that an electrostatic grid is included in the coalescing-settling section, which helps to promote coalescence of the water droplets.

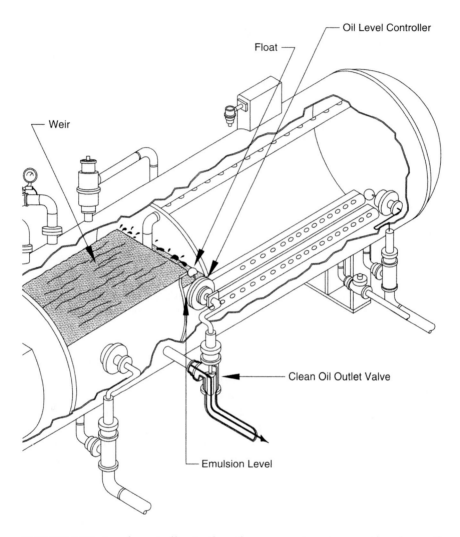

FIGURE 1.22. Level controller in the oil surge section operates the clean oil dump valve.

The electrostatic section contains two or more electrodes, one grounded to the vessel and the other suspended by insulators. An electrical system supplies an electric potential to the suspended electrode. The usual applied voltage ranges from 10,000 to 35,000 VAC, and the power consumption is from 0.05 to 0.10 kVA/ft^2 (0.54–1.08 kVA/m^2) of grid. The intensity of the electrostatic field is controlled by the applied voltage and spacing of electrodes. In some installations the location of the ground electrode can be adjusted externally to increase or decrease its spacing to the "hot" electrode. Optimum field

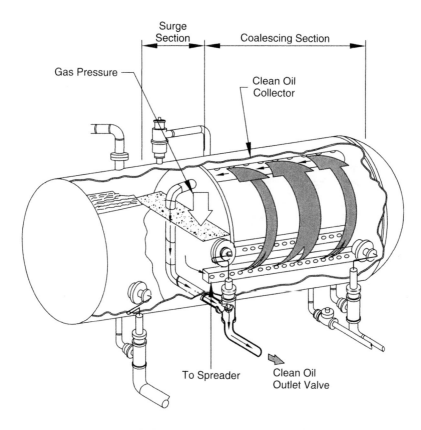

FIGURE 1.23. Pressure of the gas in the surge section forces the emulsion to flow through the spread.

intensities vary with applications but generally fall within the range of 1000–4000 V/in. (39–157 V/mm) of separation. The use of an electric field is most effective whenever the fluid viscosity is less than 50 cp at separating temperature, the specific gravity difference between the oil and water is greater than 0.001, and the electrical conductivity of the oil phase does not exceed 10^{-6} mho/cm.

The electrical control system that supplies energy to the electrodes consists of a system of step-up transformers (either single- or three-phase) in which the primary side is connected to a low-voltage power source (208, 220, or 440 V) and secondary windings are designed so that the induced voltage will be of the desired magnitude (Figure 1.26).

As shown in Figure 1.27, oil and small water droplets enter the coalescing section and travel up into the electrostatic grid section, where the water droplets become "electrified" or "ionized" and are forced to collide. The electrodes have electrical charges that reverse many times a

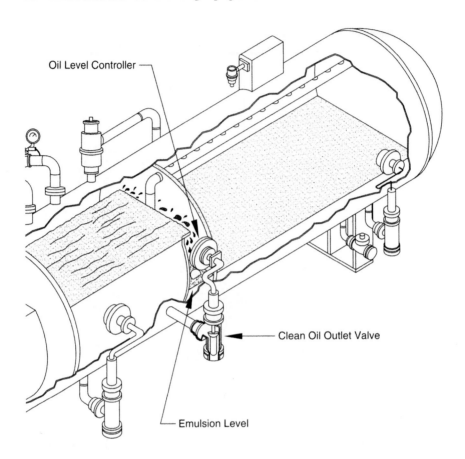

FIGURE 1.24. When the clean oil dump valve closes, the flow of emulsion to the coalescing settling section stops and the gas is prevented from entering.

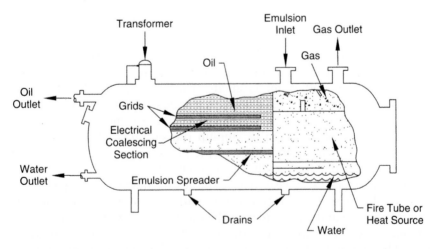

FIGURE 1.25. Simplified schematic of a horizontal electrostatic heater-treater.

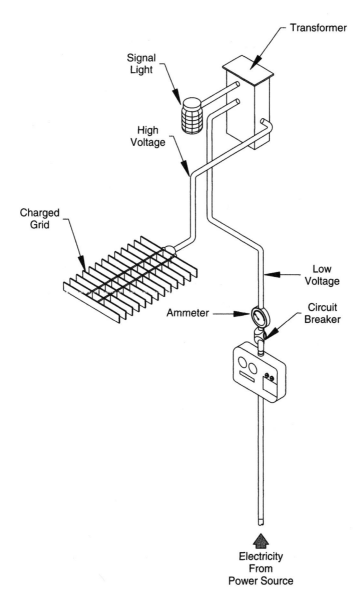

FIGURE 1.26. Electrical control system of an electrostatic heater-treater.

second; thus, the water droplets are placed in a rapid back-and-forth motion. The greater the motion of the droplets, the more likely the water droplets are to collide with each other, rupture the skin of the emulsifying agent, coalesce, and settle out of the emulsion. Because of the forced collisions, electrostatic heater-treaters typically operate at lower temperatures and use less fuel than standard heater-treaters. The time in

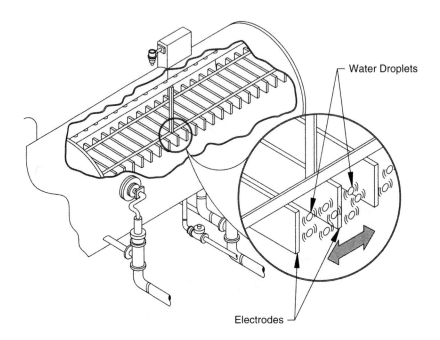

Water Droplets

Electrodes

FIGURE 1.27.   Effect of electrical charge on small water droplets in the emulsion.

the electronic field is controlled by electrode spacing and the vessel configuration. An electronic field exists throughout the body of the oil within the vessel, even though most coalescing takes place in the more intense fields in the vicinity of the electrodes.

It is imperative that the design of the vessel assure good gas removal and provide for distribution of the emulsion across the electrical grid. It is also essential to maintain the fluid in the liquid phase in the electrical coalescing section. Any vapor in the electrode area will be saturated with water. Any water-saturated vapors, which are highly conductive, will greatly increase the electrical power consumption.

It is also important to prevent the water level from reaching the height of the electrodes. Nearly all produced water contains some salt. These salts make the water a very good conductor of electric currents. Thus, if the water contacts the electrodes, it may short out the electrode grid or the transformer.

Time in the electrostatic field is controlled by electrode spacing and the vessel configuration. An electrostatic field exists throughout the body of the oil within the vessel, although most coalescing takes place in the more intense fields in the vicinity of the electrodes.

Since coalescence of the water droplets in an electric field is dependent on the characteristics of the specific emulsion being treated, sizing of grid area requires laboratory testing.

1.2.12 Oil Dehydrators

The primary factor when designing coalescing units is the loading rate. Vessels are sized for a certain volume flow per unit time per square foot of grid area. Procedures for designing electrostatic grids have not been published. Since coalescence of water droplets in an electric field is so dependent on the characteristics of the particular emulsion to be treated, it is unlikely that a general relationship of water droplet size to use in the settling equations can be developed.

Field experience tends to indicate that electrostatic treaters are effective at reducing water content in the crude to the 0.2–0.5% level. This makes them particularly attractive for oil desalting operations. However, for normal crude treating, where 0.5–1.0% BS&W is acceptable, it is recommended that the vessel be sized as a horizontal heater-treater, neglecting any contribution from the electrostatic grids. By trial and error after installation, the electric grids may be able to allow treating to occur at lower temperatures.

Figure 1.28 shows one variation of the electrostatic heater-treater where the vessel only contains the coalescing section with the electrostatic grid. Units configured in this manner are called "oil dehydrators." They are capable of higher handling volumes and use separate

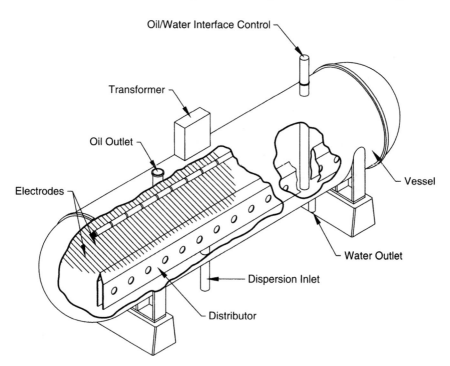

FIGURE 1.28. Cutaway of a liquid-packed horizontal oil dehydrator.

upstream vessels for free-water removal and heating. This configuration should be considered when the volume to be treated exceeds 15,000–20,000 barrels per day (BPD).

1.3 Emulsion Treating Theory

1.3.1 Introduction

Removing water from crude oil often requires additional processing beyond the normal oil–water separation process, which relies on gravity separation. Crude oil treating equipment is designed to break emulsions by coalescing the water droplets and then using gravity separation to separate the oil and water. In addition, the water droplets must have sufficient time to contact each other and coalesce. The negative buoyant forces acting on the coalesced droplets must be sufficient to enable these droplets to settle to the bottom of the treating vessel. Therefore, it is important when designing a crude oil treating system to take into account temperature, time, viscosity of the oil, which may inhibit settling, and the physical dimensions of the treating vessel, which determines the velocity at which settling must occur.

When selecting a treating system, several factors should be considered to determine the most desirable method of treating the crude oil to contract requirements. Some of these factors are:

- Stability (tightness) of the emulsion,
- Specific gravity of the oil and produced water,
- Corrosiveness of the crude oil, produced water, and associated gas,
- Scaling tendencies of the produced water,
- Quantity of fluid to be treated and percent water in the fluid,
- Paraffin-forming tendencies of the crude oil,
- Desirable operating pressures for equipment,
- Availability of a sales outlet and value of the associated gas produced.

A common method for separating this "water-in-oil" emulsion is to heat the stream. Increasing the temperature of the two immiscible liquids deactivates the emulsifying agent, allowing the dispersed water droplets to collide. As the droplets collide, they grow in size and begin to settle. If designed properly, the water will settle to the bottom of the treating vessel due to differences in specific gravity.

Laboratory analysis, in conjunction with field experience, should be the basis for specifying the configuration of treating vessels.

The purpose of this chapter is to present a rational alternative for those instances when laboratory data do not exist or, if it is desirable, to extrapolate field experience.

1.3.2 Emulsions

An emulsion is a stable mixture of oil and water that does not separate by gravity alone. In the case of a crude oil or regular emulsion, it is a dispersion of water droplets in oil. Oil is the continuous phase and water is the dispersed phase. Normal, or regular, oil-field emulsions consist of an oil continuous or external phase and a water dispersed or internal phase. In some cases, where there are high water cuts, such as when a water-drive field has almost "watered out," it is possible to form reverse emulsions with water as the continuous phase and oil droplets as the internal phase. Complex or "mixed" emulsions have been reported in low-gravity, viscous crude oil. These mixed emulsions contain a water external phase and have an internal water phase mixed in the oil dispersed phase. A stable or "tight" emulsion occurs when the water droplets will not settle out of the oil phase due to their small size and surface tension. Stable emulsions always require some form of treatment. The vast majority of oil treating systems deal with normal emulsions, which is the focus of this chapter.

For an emulsion to exist there must be two mutually immiscible liquids, an emulsifying agent (stabilizer), and sufficient agitation to disperse the discontinuous phase into the continuous phase. In oil production, oil and water are the two mutually immiscible liquids. When oil and water are produced from a well, the fluid stream also contains organic and inorganic materials. These contaminants are preferentially absorbed at the interface between the oil and water phases. Once the contaminants are absorbed at the interface, they form a tough film (skin) that impedes or prevents the coalescence of water droplets. The film prevents the water droplets from coalescing. Agitation, sufficient to disperse one liquid as fine droplets through the other, occurs as the well fluids makes their way into the well bore, up the tubing, and through surface chokes, down-hole pumps, and gas lift valves. Turbulence caused by the pressure drop across the choke is the primary source of agitation for emulsion formation. However, elimination of the choke, used to control the flow rate of a well, is not a solution to the problem.

The degree of agitation and the nature and amount of emulsifying agent determine the "stability" of the emulsion. Some stable emulsions may take weeks or months to separate if left alone in a tank with no treating. Other unstable emulsions may separate into relatively pure oil and water phases in just a matter of minutes. The stability of an emulsion is dependent on several factors:

- the difference in density between the water and oil phases,
- the size of dispersed water particles,
- viscosity,
- interfacial tension,
- the presence and concentration of emulsifying agents,
- water salinity,
- age of the emulsion, and
- agitation.

1.3.3 Differential Density

The difference in density between the oil and water phases is one of the factors that determines the rate at which water droplets settle through the continuous oil phase. The greater the difference in gravity, the more quickly the water droplets will settle through the oil phase. Heavy oils (high specific gravity) tend to keep water droplets in suspension longer. Light oils (low specific gravity) tend to allow water droplets to settle to the bottom of the tank. Thus, the greater the difference in density between the oil and water phases, the easier the water droplets will settle.

1.3.4 Size of Water Droplets

The size of the dispersed water droplets also affects the rate at which water droplets move through the oil phase. The larger the droplet, the faster it will settle out of the oil phase. The water droplet size in an emulsion is dependent upon the degree of agitation that the emulsion is subjected to before treating. Flow through pumps, chokes, valves, and other surface equipment will decrease water droplet sizes.

1.3.5 Viscosity

Viscosity plays two primary roles in the stability of an emulsion. First, as oil viscosity increases, the migration of emulsifying agents to the water droplet's oil–water interface is retarded. This results in larger water droplets being suspended in the oil, which in turn results in less stable emulsions in terms of numbers of small water droplets suspended in the oil. As oil viscosity increases, more agitation is required to shear the larger water droplets down to a smaller size in the oil phase. Thus, the size of the water droplets that must be removed to meet water cut specifications for a given treating system increases as viscosity increases. Second, as viscosity increases, the rate at which water droplets move through the oil phase decreases, resulting in less coalescence and increased difficulty in treating. On the other hand, as oil viscosity decreases, the friction encountered by the water droplets moving through the continuous oil phase is reduced, which in turn promotes separation of the oil and water phases.

1.3.6 Interfacial Tension

Interfacial tension is the force that "holds together" the surfaces of the water and oil phases. When an emulsifying agent is not present, the interfacial tension between oil and water is low. When interfacial tension is low, water droplets coalesce easily upon contact. However, when emulsifying agents are present, they increase the interfacial tension and obstruct the coalescence of water droplets. Anything that lowers the interfacial tension will aid in separation.

1.3.7 Presence and Concentration of Emulsifying Agents

Chemicals (demulsifiers) are normally used to reduce the interfacial tension. Chemical effectiveness is enhanced by mixing, time, and temperature. Adequate mixing and sufficient time are required to obtain intimate contact of the chemical with the dispersed phase. A certain minimum temperature is required to ensure the chemical accomplishes its function. Both viscosity reduction and effectiveness of chemical are dependent on the attainment of a certain minimum temperature. It may well be that the increase in chemical effectiveness is a result of the decrease in viscosity of the oil phase.

1.3.8 Water Salinity

The salinity of the water is a measure of the total dissolved solids in the water phase. As salinity of the water increases, the density of the water increases, which in turn increases the differential density between the water and the oil. The increase in differential density aids in separation of the oil and water phases. Small amounts of salt, or other dissolved solids, in the water phase will appreciably lower the interfacial tension and thus will decrease the difficulty of separating the two phases. To some degree, this phenomenon explains the difficulty of treating water–oil emulsions formed from soft water typically found in many steam flood operations, for example, Caltex Duri Field, Sumatra, Indonesia, and Chevron Texaco, Bakersfield, California.

1.3.9 Age of the Emulsion

As emulsions age they become more stable and separation of the water droplets becomes more difficult. The time required to increase stability varies widely and depends on many factors. Internal and external properties of the stream will change throughout the life of production due to changes in formation characteristics and fluctuations in the ambient conditions encountered on the surface. This partially explains the ever-changing problems associated with emulsion treating. Little or no emulsion exists in oil bearing formations. Emulsions are formed during production on the fluid. The degree of emulsification is dependent on the agitation of the two phases by pumps, chokes, etc. Before an emulsion is produced, the emulsifying agents are evenly dispersed in the oil. As soon as the water phase is mixed with the oil, the emulsifying agents begin to cluster around the water droplet to form a stable emulsion. While the initial stabilization may occur in a matter of a few seconds, the process of film development may continue for several hours. It will continue until the film around the droplet of water is so dense that no additional stabilizer can be attracted, or until there no stabilizer is left to be extracted from the oil. At such a time the emulsion has reached a state of equilibrium and is said to be aged. The older the emulsion, the more difficult it is to treat. Therefore, emulsion breaking or treating operations are often located as close to the wellhead as possible, so that emulsions formed during flow in the production tubing and wellhead equipment are not allowed to age before treatment.

1.3.10 Agitation

The type and severity of agitation applied to an oil–water mixture determine the water drop size. The more turbulence and shearing action present in a production system, the smaller the water droplets and the more stable the emulsion will be.

The above factors determine the "stability" of emulsions. Some stable emulsions may take weeks or months to separate if left alone in a tank with no treating. Other unstable emulsions may separate into relatively clean oil and water phases in just a matter of minutes.

Figure 1.29 shows a normal water-in-oil emulsion. The small water droplets exist within the oil continuous phase. Figure 1.30 shows a close-up of a "skin" (monomolecular film) of emulsifying agent surrounding a water drop, and Figure 1.31 shows two drops touching but unable to coalesce because of the emulsifying-agent "skin" surrounding each drop.

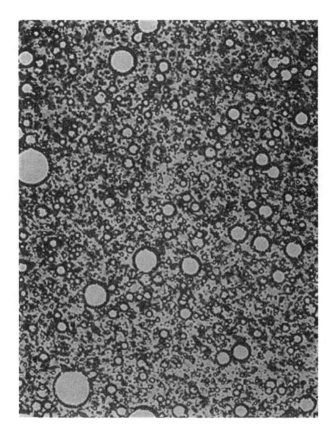

FIGURE 1.29. Photomicrograph of an water-in-oil emulsion.

1.3.11 Emulsifying Agents

When studying emulsion stability, it may be helpful to realize that in a pure oil and pure water mixture, without an emulsifying agent, no amount of agitation will create an emulsion. If the pure oil and water are mixed and placed in a container, they quickly separate. The natural state is for the immiscible liquids to establish the least contact or smallest surface area. The water dispersed in the oil forms spherical drops. Smaller drops will coalesce into larger drops, and this will create a smaller interface area for a given volume. If no emulsifier is present, the droplets will eventually settle to the bottom, causing the smallest interface area. This type of mixture is a true "dispersion."

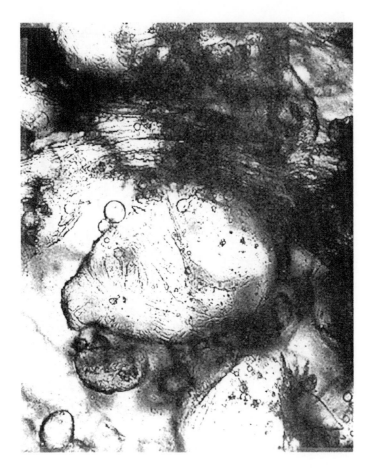

FIGURE 1.30. Photomicrograph showing a close-up view of the emulsifying agent skin surrounding a water droplet.

An emulsifying agent in the system is a material, which has a surface-active behavior. Some elements in emulsifiers have a preference for the oil, and other elements are more attracted to the water. An emulsifier tends to be insoluble in one of the liquid phases. It thus concentrates at the interface. There are several ways emulsifiers work to cause a dispersion to become an emulsion. The action of the emulsifier can be visualized as one or more of the following:

- It forms a viscous coating on the droplets, which keeps them from coalescing into larger droplets when they collide. Since coalescence is prevented, it takes longer for the small droplets, which are caused by agitation in the system, to settle out.

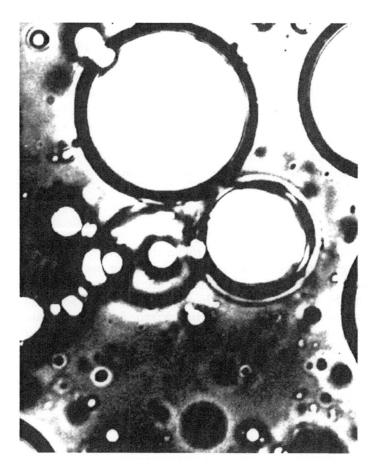

FIGURE 1.31. Photomicrograph showing two droplets touching but unable to coalesce because of the emulsifying skin surrounding the droplets.

- The emulsifiers may be polar molecules, which align themselves in such a manner as to cause an electrical charge on the surface of the droplets. Since like electrical charges repel, two droplets must collide with sufficient force to overcome this repulsion before coalescence can occur.

Naturally occurring surface-active materials normally found in crude oil serve as emulsifiers. Paraffins, resins, organic acids, metallic salts, colloidal silts and clay, and asphaltenes (a general term for material with chemical compositions containing sulfur, nitrogen, and oxygen) are common emulsifiers in oil fields. Workover fluids and drilling mud are also sources of emulsifying agents.

The type and amount of emulsifying agent have an immediate effect on the emulsion's stability. It has been shown that the temperature history of the emulsion is also important as it affects the formation of paraffins and asphaltenes. The speed of migration of the emulsifying agent to the oil–water interface and the behavior in terms of the strength of the interface bond are important factors. An emulsion treated soon after agitation, or soon after the creation of paraffins and asphaltenes, can be less stable and easier to process if the migration of the emulsifier is incomplete. An aged emulsion may become more difficult to treat because the emulsifying agents have migrated to the oil–water interface. Normally, the lower the crude viscosity and lighter the crude, the more rapid the aging process will be.

1.3.12 Demulsifiers

Emulsions can be resolved or broken thermally and/or chemically. When we chemically resolve an emulsion, we use a demulsifier or emulsion breaker. These two names are used interchangeably and describe the same chemical. Chemical demulsifiers sold under various trade names, such as Tretolite, Visco, Breaxit, etc., are highly useful in resolving emulsions. Demulsifiers act to neutralize the effect of emulsifying agents. Typically, they are surface-active agents and thus their excessive use can decrease the surface tension of water droplets and actually create more stable emulsions. In addition, demulsifiers for water-in-oil emulsions tend to promote oil-in-water emulsions; therefore, excessive chemical use may cause water treating problems.

Four important actions are required of a demulsifier:

- strong attraction to the oil–water interface,
- flocculation,
- coalescence, and
- solid wetting.

When these actions are present, they promote the separation of oil and water. The demulsifier must have the ability to migrate rapidly through the oil phase to the droplet interface, where it must compete with the more concentrated emulsifying agent. The demulsifier must produce an attraction for similar droplets. In this way large clusters of droplets gather, which, under a microscope, appear like bunches of fish eggs. The oil will take on a bright appearance since small droplets are no longer present to scatter the light rays. At this point the emulsifier film is still continuous. If the emulsifier is weak, the flocculation force may be enough to cause coalescence. This is not true in most cases, and the demulsifier must therefore neutralize the emulsifier and promote a rupture of the droplet interface film. This is the opener that causes coalescence. With the emulsion in a flocculated condition, the film rupture results in rapid growth of water drop size.

The manner in which the demulsifier neutralizes the emulsifier depends upon the type of emulsifiers. Iron sulfides, clays, and drilling muds can be water wet, causing them to leave the interface and be diffused into the water droplet. Paraffins and asphaltenes could be dissolved or altered to make their films less viscous so they will flow out of the way on collision or could be made oil wet so they will be dispersed in the oil.

It would be unusual if one chemical structure could produce all four desirable actions. A blend of compounds is therefore used to achieve the right balance of activity.

The demulsifier selection should be made with the process system in mind. If the treating process is a settling tank, a relatively slow-acting compound can be applied with good results. On the other hand, if the system is a chemical–electric process where some of the flocculation and coalescing action is accomplished by an electric field, there is need for a quick-acting compound, but not one that must complete the droplet-building action.

As field conditions change, the chemical requirements can change. If the process is modified, for example, very low rates on electrostatic units, the chemical requirements can change. Seasonal changes bring paraffin-induced emulsion problems. Workovers contribute to solids and acid/base contents, which alters the emulsion stability. So no matter how satisfactory a demulsifier is at one point in time, it may not be satisfactory over the life of the field.

The cost to dehydrate crude oil chemically is a function of several factors. First, the ratio of oil to water is important—it is generally easier and, hence, less costly to dehydrate crudes with high water cuts. Next, the severity of the emulsion is important. A "tight" emulsion consisting of small droplets is much more difficult to break—it has a higher surface area to volume ratio than a "loose" emulsion and, hence, the demulsifier has more work to do to seek out the interface. Next, the residence time available for separation is important. Small residence times inhibit complete separation of water droplets from oil. This may lead to re-entrainment of water as the crude goes from one processing stage to another. The result is ineffective dehydration. Higher temperatures result in lower oil phase viscosities, which enable the demulsifier to migrate to the oil–water interface faster and for coalesced water droplets to drop out easier.

Last, the dehydration cost is directly influenced by chemical selection. Poor chemical selection will result in a non-optimized treatment, which will mean higher costs. Chemical selection is not a simple process—it is best left to suppliers. However, one can assist in the process by providing on-site testing opportunities for chemical suppliers to select the best chemicals for specific applications.

1.3.13 Bottle Test

This is one of the most common, yet least understood, of all the chemical selection tests. Emulsion-breaking chemicals are most commonly tested with a bottle test, which involves mixing various chemicals with samples of the emulsion and observing the results. Such tests are effective in eliminating some chemicals and selecting those that appear to be more efficient. Bottle tests also provide an estimate of the amount of chemical required and an estimate of the settling time required for a treating vessel. Bottle tests should be performed on a representative sample as soon as the sample is obtained because of the possible detrimental effects of aging. These tests should be performed at conditions that are as close to field treating conditions as possible. Synthetic water should not be used in place of produced water in bottle tests because the produced water may have very different properties, and it may contain impurities that are not present in the synthetic water.

While candidate chemicals and approximate dosages can be determined in bottle tests, the dynamic nature of the actual flowing system requires that several candidates be field-tested. In actual conditions, the emulsion undergoes shearing through control valves, coalescence in flow-through pipes, and changes to the emulsion that occur inside the treating vessel as a result of inlet diverters, water-wash sections, etc. Static bottle tests cannot model these dynamic conditions.

As field conditions change, for example, emulsifying agents change or saltwater percentages change, the chemical requirements can change. If the process is modified, for example, very low rates on electronic units, the chemical requirement can change. Seasonal changes bring paraffin-induced emulsion problems. Workovers contribute to solids content, which alters emulsion stability. So no matter how satisfactory a demulsifier is at one point in time, it cannot be assumed that it will always be satisfactory over the life of the field.

As well as determining the potential dehydration performance of a demulsifier, the bottle test can also be used to investigate chemical incompatibilities. Here, the performance of a demulsifier is evaluated on a chemical-free sample and then on a sample of crude, which includes the other production chemicals at their respective dose rates. The change in performance, if any, is recorded and the chemical discarded if incompatibilities exist. Another aspect of incompatibility may also be determined, namely, in which order the chemicals should be injected. If the bottle tester is experienced, this order of injection, which will produce subtle changes in the bottle test results, can be investigated and an optimum injection order determined.

1.3.14 Field Trial

Having selected a promising demulsifier candidate, a field trial should be carried out to test the chemical's ability to operate in a dynamic system. In the field test, the flexibility of the demulsifier to process changes can be established. This data will be useful when the chemical is used in full-scale operation. In most field trial situations, the demulsifier being tested is first used in conjunction with a test separator system. This enables the supplier to look at the response of the chemical to one or more wells and to provide the tester an idea of the true field dosage. If this preliminary scenario is successful, the chemical can then be dosed into the full system and optimized for different well configuration and flow rates. In the field trial, the chemical's response to system upsets can be determined and, hence, an operating response can be set.

1.3.15 Field Optimization

After a successful field trial, a full-scale field optimization is carried out. Here, the chemical performance is monitored routinely as are the possible side effects of under- or overdosing, such as separator interface build-up. It may be that if the field produces through two or more platforms, injection locations and dose rates may need to be optimized for each location.

1.3.16 Changing the Demulsifier

As crude characteristics change over the life of a field, the performance of the demulsifier chemical will change also. Typically, when fields first produce water, the emulsions formed are difficult to break. As the field ages and the water cut increases, the stability of the emulsion and even the emulsifying agents themselves may change. Hence, it is usual to investigate demulsifier performance every 2–3 years. In some cases where a step change in water cut is experienced, it may be prudent to investigate demulsifier performance more frequently. In most cases a quick bottle test is all that is required to determine if the current chemical is still optimum. If not, a full bottle test to find a more effective chemical can be undertaken.

1.3.17 Demulsifier Troubleshooting

The most common problem with demulsifiers is overdosing. Poor treatment, dirty water, and interface pad build-up are all symptoms of overdosing an optimum chemical. Overdosing can occur by a step increase in dose rate, for example, going from 5–20 ppm, or by a gradual accumulation of chemical in the system. The latter is most often seen in high water cut systems where a small change from optimum can result in dirty water. The gradual accumulation of chemical usually occurs at the separator interface and is often difficult to detect. However, highly variable water quality caused by intermittent interface sloughing is often a clue to this scenario.

Other problems with demulsifiers can be that their viscosity changes with temperature. Most demulsifiers are viscous chemicals whose ability to be pumped can drop dramatically with reduced temperature. If this is the case, it may be prudent to ask the chemical supplier to produce a "winterized" version of the chemical. This is often done by reducing the percentage of active ingredient and adding a more solvent carrier. If this is the solution, the dose rates will need to be re-optimized for best performance.

Another common problem with demulsifiers is their apparent lack of treatment "range." It is not uncommon for a field demulsifier to have a different performance standard for different wells within a field. In some cases "rogue wells" may exist, which are basically untreatable by the optimum demulsifier for the rest of the system. In these cases two demulsifiers may be used or the original demulsifier may be injected at a higher dose rate or even down-hole in the rogue well. The bottle test will often indicate rogue wells and their best treatment solution.

Demulsifiers and corrosion inhibitors are often the cause of poor dehydration performance. Corrosion inhibitors are surfactant chemicals that often act as emulsifying agents, thus making the demulsifier work harder. In cases of conflict, it is usually easier to blend a new demulsifier or change the injection points of the chemicals than it is to replace the corrosion inhibitor. However, in some North Sea fields the opposite was true. Corrosion inhibitor replacement was the best way to deal with the incompatibility problem.

As there are no online analyzers for demulsifier performance, one must monitor the facilities for changes in water or crude quality that may be attributed to poor demulsifier performance. Chemical suppliers can help here by giving us the anticipated system response to incompatibilities and over- or under-dosing. They should get this information from the bottle test and the demulsifier field trial.

1.4 Emulsion Treating Methods

1.4.1 General Considerations

Treating processes and equipment should not be selected until the physical characteristics of the oil and water have been determined and a study of the effect of available chemicals on the emulsion has been made.

The water remaining in the crude after the free water has settled out is considered to be in an emulsified state. Emulsified oil is removed by one or more treating processes. Treating refers to any process designed to separate crude oil from water and foreign contaminants carried along with it from the reservoir. Emulsion treating processes require some combination of the following: chemical addition, settling time, heat, and electrostatic coalescing.

1.4.2 Chemical Addition

The purpose of treating chemicals is to induce coalescence so that the oil and water will separate rapidly. Surface-active agents are absorbed at the oil–water interface, rupture the tough film (skin) surrounding the water droplets, and/or displace the emulsifying agent and force the emulsifying agent back into the oil phase.

There is no universal chemical that will break all emulsions equally well. Determining the correct chemical to use is commonly done by a chemical sales representative using a bottle test (discussed earlier in this chapter).

1.4.3 Amount of Chemical

The amount of chemical required cannot be predicted accurately from bottle tests. The only reliable method of determining the amount of chemical to use is to run tests in the field. When changing to a new chemical or starting up a new treating system, one must first use an excess (1 quart per 100 barrels) of chemical and then gradually reduce the amount to the minimum amount that will produce the desired results. When determining the amount of chemical to add, one must make certain no other changes are being made in the facility. Temperature should remain constant during the test; otherwise, it is impossible to determine which change, chemical or temperature, has caused a certain effect.

The amount of chemical added can vary from 1 gallon per 150 to 1000 barrels. Concentrations higher than 1 gallon per 250 barrels should be investigated for possible errors such as incorrect chemical being used or the method of chemical addition being wrong. Too much chemical can be the cause of a very tight emulsion that will not break down.

Chemicals should be added continuously as possible during the entire production period and at a rate related to the production rate. Even though some residual chemical is held in the treater or gunbarrel, chemicals cannot be batched and be expected to do an adequate treatment. Chemicals cannot act properly unless they are thoroughly mixed with the emulsion. The farther upstream, a minimum of 200 ft, from a treater or gunbarrel the chemicals are added results in better mixing and better treatment. The ideal location for injection is at the manifold before the fluid enters a separator. In some cases an emulsion that is difficult to treat may break quite easily if a chemical pump is set at the well. It is not uncommon for one well in a field to cause most of the trouble. Setting a pump at this well can increase efficiency and reduce the amount of chemicals required to break the emulsion.

1.4.4 Bottle Test Considerations

The best demulsifier is the compound that results in the most rapid and complete separation of the phases at a minimum concentration. The important characteristics in the bottle test will be dictated by the production needs and the behavior of the system.

Water Drop-Out Rate

In high water volume systems a chemical that creates a fast water drop-out rate is necessary to make the system function as designed. When FWKOs are used, the speed of water drop-out may become the most important factor. Chemicals with fast water drop-out characteristics are sometimes incomplete in treatment. In low water volume systems (fields with facilities having longer than normal residence times) the rate of water drop-out may be of minor significance in selecting the best demulsifier. In all cases, the rate of water drop-out should be noted and recorded.

Sludge

When sediment and water collect without breaking to water and oil, the result is called sludge. In some systems, non-coalesced water droplets result in a loose agglomeration that breaks to water and oil, causing no problems. Depending upon the system and sludge stability, interface sludge may or may not cause a problem. Sludge is stabilized by finely divided solids and other contaminates to form pads that cause a secondary emulsion located between the oil and water. Loose interface sludge can be detected by swirling the test bottle about its axis, and if the material is loose, it will break.

Interface

The desired interface is one that has shiny oil in contact with the water (mirror interface). The interface, when using a new chemical, should be as good as, if not better than, that formed by the chemical being replaced.

Water Turbidity

The turbidity (clarity) of the water is very difficult to interpret in the bottle test and correlate to facility behavior. When the chemical effects in the bottle are pronounced and reproducible, some correlation can be expected. Clear water is definitely the desired result.

Oil Color

Emulsions have a hazy appearance when compared to the bright color of treated oil. As a crude oil emulsion separates, the color tends to brighten. Brightening of oil can be encouraging, but it can also be deceptive if taken as the sole qualification for chemical selection. While bright color is no guarantee of a successful chemical, lack of it assures that the compound is not worthy of further consideration.

Centrifuge Results

An important quality in the final evaluation is the centrifuge results. It is always good practice to make a centrifuge grind out to accurately determine the final amount of BS&W entrained in the oil.

1.4.5 Chemical Selection

A thorough understanding of the treating equipment and its contribution to the treatment are necessary before chemical selection can be made. If little agitation is available, a fast-acting chemical is necessary. If an FWKO vessel is used, the water drop-out rate will be very important. If heat is unavailable, the chemical must work at ambient temperatures. Different types of vessels require different chemical actions.

Settling Tank or "Gunbarrel"

Speed is not too important since both tanks usually have a high volume-to-throughput ratio. The chemical may continue acting over a relatively long period. An interface layer often develops but usually stabilizes at some acceptable thickness. An interface layer in a gunbarrel sometimes aids the treating process in that it acts as a filter for solids and unresolved emulsions. Fresh oil containing a demulsifier passing up through the interface layer helps treat the interface and prevents an excessive build-up.

Vertical Heater-Treater

The speed of chemical action is important since the volume-to-throughput ratio is usually lower than a gunbarrel or settling tank.

With the higher throughput, it is harder to stabilize an interface layer, so more complete treatment is necessary in a shorter time period. Solids control may be important in controlling the interface.

Horizontal Heater-Treater

The speed of chemical action is important due to its high throughput. The large interface area and shallow depth require that the interface be fairly clean. Since this treater can tolerate only very little interface accumulation, the chemical treatment must be complete. Since solids tend to collect at the interface, the chemical must also effectively de-oil any solids so that they may settle out by gravity.

1.4.6 Settling Time

Following the addition of treating chemicals, settling time is required to promote gravity settling of the coalescing water droplets. Figure 1.32 illustrates the effects of time on coalescence. Emulsion-treating equipment designed to provide sufficient time for free water to settle includes three-phase separators, FWKOs, heater-treaters, and gunbarrels with an internal or external gas boot. The time necessary for free

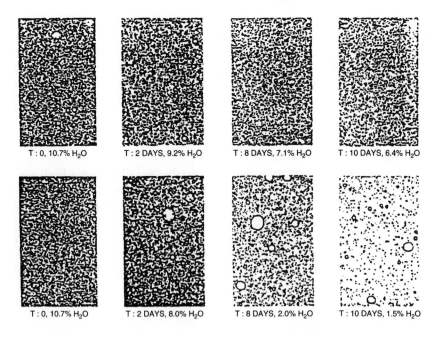

| T : 0, 10.7% H_2O | T : 2 DAYS, 9.2% H_2O | T : 8 DAYS, 7.1% H_2O | T : 10 DAYS, 6.4% H_2O |

| T : 0, 10.7% H_2O | T : 2 DAYS, 8.0% H_2O | T : 8 DAYS, 2.0% H_2O | T : 10 DAYS, 1.5% H_2O |

FIGURE 1.32. Effect of time on coalescence. *Top*: emulsion without chemicals. *Bottom*: emulsion with demulsifier added.

water to settle is affected by differential density of the oil and water, viscosity of the oil, size of the water droplets, and relative stability of the emulsion.

1.4.7 Coalescence

The process of coalescence in oil treating systems is time-dependent. In dispersions of two immiscible liquids, immediate coalescence seldom occurs when two droplets collide. If the droplet pair is exposed to turbulent pressure fluctuations, and the kinetic energy of the oscillations induced in the coalescing droplet pair is larger than the energy of adhesion between them, the contact will be broken before coalescence is completed.

Experiments with deep-bed gravity settlers indicate that the time to 'grow' a droplet size due to coalescence can be estimated by the following equation:

$$t = \frac{\pi}{6}\left(\frac{d^j - (d_o)^j}{\psi K_s}\right), \tag{1.1}$$

where

d_o = initial droplet size, μm,
d = final droplet size, μm,
ψ = volume fraction of the dispersed phase,
K_S = empirical parameter for the particular system,
j = an empirical parameter that is always larger than 3 and dependent on the probability that the droplets will "bounce" apart before coalescence occurs,
t = time required to grow a droplet of size d, min.

When the energy of oscillations is very low so that "bouncing" of droplets approaches 0, j approaches 3. Assuming a value of 4, the minimum time required to obtain a desired particle diameter can be expressed as

$$t = \frac{\pi}{6}\left(\frac{d^4 - (d_o)^4}{\psi K_s}\right). \tag{1.2}$$

Assuming d_o is small relative to the droplet size we wish to "grow" by coalescence in our gravity settler, Equation (1.2) can be approximated:

$$t = \frac{d^4}{2\psi K_s}. \tag{1.3}$$

The following qualitative conclusions for coalescence in a gravity settler can be drawn from this relationship:

- A doubling of residence time increases the maximum size drop grown in a gravity settler less than 19%. If $j > 4$, the growth in droplet diameter will be even slower. For this reason, after an initial short coalescence period, adding additional retention time is not very effective for making the oil easier to treat. Very often engineers will attribute improved performance in large gunbarrel tanks to retention time when it is really due to slowing the oil velocity. This allows smaller droplets of water to separate in accordance with Stokes' law.
- The more dilute the dispersed phase, the greater the residence time needed to "grow" a given particle size will be. That is, coalescence occurs more rapidly in concentrated dispersions. This is the reason that oil is "water-washed" by entering the treating vessel below the oil–water interface in most gunbarrels and treaters. Flocculation and coalescence, therefore, occur most effectively at the interface zone between oil and water.

1.4.8 Viscosity

The viscosity of the oil continuous phase is extremely important in sizing a treater. Stokes' law, used to determine the settling velocity of a water droplet settling through the continuous oil phase, includes the oil viscosity. As the oil viscosity increases, the settling velocity of a given droplet decreases. This requires that the treater size be increased.

The oil viscosity also affects coalescence of the water droplets. As the oil viscosity increases, there is more resistance to random motion of the water droplets. Therefore, the droplets do not move as fast or as far. This decreases the energy and the frequency of droplet collisions. Thus, it is more difficult to grow large water droplets in the vessel. As the oil viscosity increases, it is also more difficult to shear the oil droplets that coalesce in the piping leading to the vessel and in the water-wash section of the vessel. The net effect is that increasing the oil viscosity increases the size of the minimum water droplet that must be removed.

By far the best situation is to have oil viscosity versus temperature data for a particular oil to be treated. Alternately, data from other wells in the same field can usually be used without significant error. This viscosity versus temperature data may be plotted on special ASTM graph paper. Such plots are usually straight lines, unless the oil has a high cloud point. The viscosity may then be predicted at any temperature.

Laboratory testing of a particular oil at various temperatures is the most reliable method of determining how an oil behaves. ASTM

D-341 outlines a procedure where the viscosity is measured at two different temperatures and then, either through a computation or on special graph paper, the viscosity at any other temperature can be obtained.

As a rule, with crude of 30 °API and higher, the viscosity is so low that normally it may be difficult to find any information on file regarding a specific crude viscosity. Between 30 and 11 °API, the viscosity becomes more important, until in some cases it is impossible to process very low gravity crudes without a diluent to reduce the viscosity. The use of a diluent is not unusual for crude oil below 14 °API.

With virtually any crude oil the viscosity change with temperatures can be an excellent guide to minimum crude processing temperatures. An ASTM chart of the viscosity versus temperature is useful to detect the paraffin formation or cloud point of the crude as shown in Figure 1.33. This normally establishes a minimum temperature for the treating process. There are examples of 30 °API crude and higher that have pour-points of 80–90 °F (27–32 °C). Crude oils of this type are common in the Uinta and Green River Basins of the United States as well as in Southeast Asia.

If no data are available, the oil viscosity may be estimated by a variety of methods from the temperature and oil gravity. These methods, however, are not very accurate, as the viscosity is a function of the oil composition and not strictly the oil gravity. In fact, two oils with the same gravity at the same temperature may have viscosities that are orders of magnitude apart.

In the absence of any laboratory data, Figures 1.34 and 1.35 may be used to estimate oil viscosities. Additional correlations that can be used to estimate crude viscosity given its gravity and temperature are discussed in the *Gas–Liquid and Liquid–Liquid Separation* volume.

1.4.9 Heat Effects

Adding heat to the incoming oil–water stream is the traditional method of separating the phases. The addition of heat reduces the viscosity of the oil phase, allowing more rapid settling velocities in accordance with Stokes' law of settling. For some emulsifying agents, such as paraffins and asphaltenes, the addition of heat deactivates, or dissolves, the emulsifier and thus increases its solubility in the oil phase. Treating temperatures normally range from 100 to 160 °F (38–70 °C). In treating of heavy crudes the temperature may be as high as 300 °F (150 °C).

Adding heat can cause a significant loss of the lower-boiling-point hydrocarbons (light ends). This results in "shrinkage" of the oil or loss of volume. The molecules leaving the oil phase may be used as fuel, vented, or compressed and sold with the gas. Even if they are

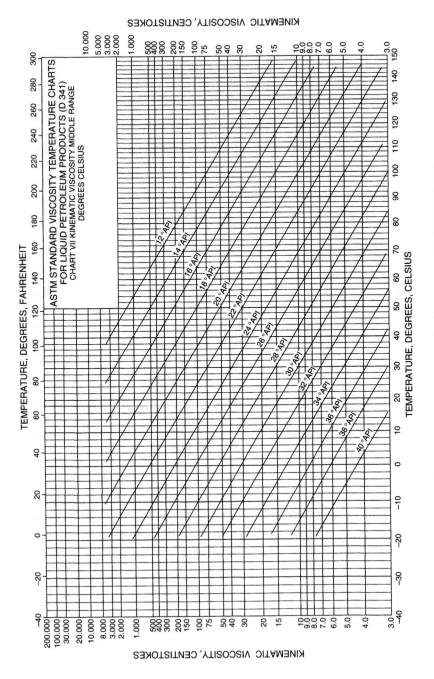

FIGURE 1.33. Viscosity versus temperature for several crude oils. (Courtesy of ASTM D-341.)

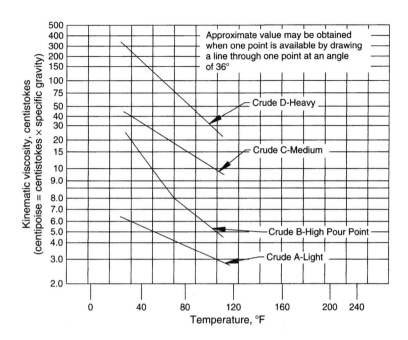

FIGURE 1.34. Typical oil viscosity versus temperature and gravity for estimating purposes, field units.

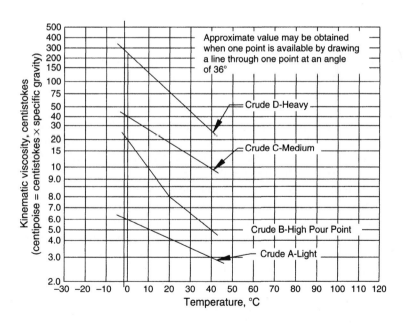

FIGURE 1.35. Typical oil viscosity versus temperature and gravity for estimating purposes, SI units.

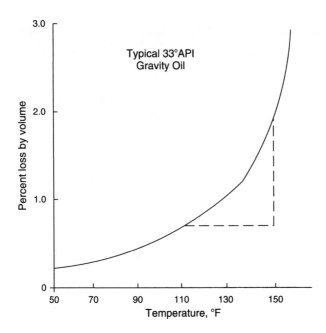

FIGURE 1.36. Percent loss by volume as a function of temperature for a 33 °API gravity crude oil.

sold with the gas, there will probably be a net loss in income realized by converting liquid volume into gas volume. Figure 1.36 shows the amount of shrinkage that may be expected from a typical 33 °API gravity crude oil.

Increasing the temperature at which treating occurs has the disadvantage of making the crude oil that is recovered in the storage tank heavier and thus decreasing its value. Because the light ends are boiled off, the remaining liquid has a lower API gravity. Figure 1.37 shows the API gravity loss for a typical crude oil.

Increasing the temperature may lower the specific gravity, at the treater operating pressure, of both the oil to be treated and the water that must be separated from it. However, depending on the properties of the crude, it may either increase or decrease the difference in specific gravity as shown in Figure 1.38.

In most cases, if the treating temperature is less than 200 °F (93 °C), the difference between the oil and water specific gravities (ΔSG) is constant and thus can be neglected. The variation of oil specific gravity with temperature is approximately as shown in Figure 1.39 adapted from the *GPSA Engineering Data Book*. The specific gravity for water is given in Figure 1.40.

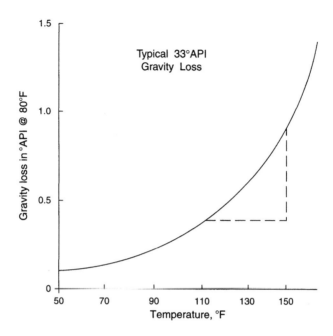

FIGURE 1.37. API gravity loss as a function of temperature for a 33 °API gravity crude oil.

Finally, it takes fuel to provide heat, and the cost of fuel must be considered. Thus, while heat may be needed to treat the crude adequately, the less heat that is used, the better. Using data from a 1983 study, Table 1.4 illustrates the overall economic effect of treating temperature for a lease that produces 21,000 BOPD (139 m³/h) of a 29 °API crude.

The gas liberated when crude oil is heated may create a problem in the treating equipment if the equipment is not properly designed. In vertical heater-treaters and gunbarrels the gas rises through the coalescing section. If much gas is liberated, it can create enough turbulence and disturbance to inhibit coalescence. Perhaps more important is the fact that the small gas bubbles have an attraction for surface-active material and hence for the water droplets. The bubbles thus have a tendency to keep the water droplets from settling and may even cause them to carry over to the oil outlet.

The usual oil-field horizontal heater-treater tends to overcome the gas liberation problem by coming to equilibrium in the heating section before introducing the emulsion to the coalescing-settling section. Some large crude processing systems use a fluid-packed, pump-through system that keeps the crude well above the bubble-point. Top-mount degassing separators above electrostatic coalescers have been used in some installations.

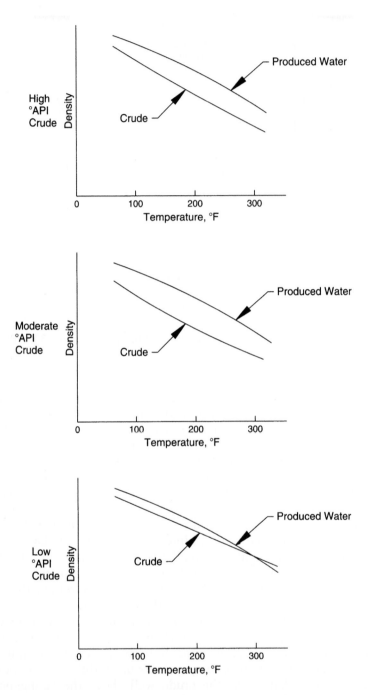

FIGURE 1.38. Relationship of specific gravity and temperature for a high, moderate, and low API gravity crude.

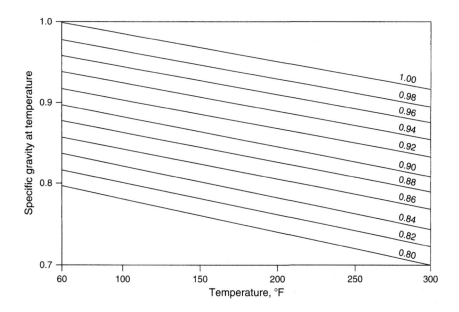

FIGURE 1.39. Variation of specific gravity of petroleum fractions with temperature.

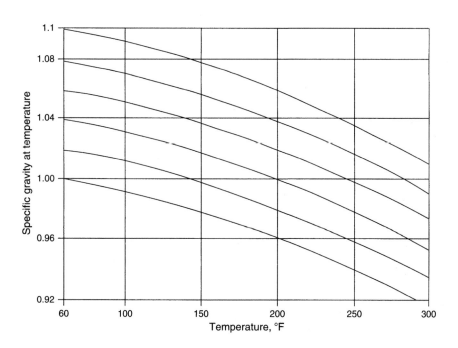

FIGURE 1.40. Variation of specific gravity of water with temperature.

TABLE 1.4
Economic effect of treating at a higher temperature for a specific field

Increase in NGL Value: Component	Volume for 120 °F Treater Temperature	Volume for 100 °F Treater Temperature	Difference	Price Per Unit[a]	Change NGL Rev (/h)
Methane	163 mcfh	162 mcfh	1 mcfh	$2.44	2.44
Ethane	1835 gal/h	1802 gal/h	33 gal/h	0.232	7.66
Propane	1653 gal/h	1527 gal/h	126 gal/h	0.439	55.31
Butane	1086 gal/h	930 gal/h	156 gal/h	0.672	104.83
Pentane+	1251 gal/h	968 gal/h	283 gal/h	0.749	211.97
					382.21

	$/Day
Total NGL revenue gain	$9173.04
Net value to producer	$3057.68
Volume shrinkage	
317 BOPD × $19.52/bbl	(6187.84)
API gravity loss	
20,931 BPD × $0.15/bbl × 7/10 (API change)	(2197.75)
Fuel cost	
125 Mcfd × $2.44/Mcf =	(306.00)
Total lease revenue loss:	$5633.91

Source: V. L. Heiman et al.: "Maximize Revenue by Analyzing Crude Oil Treating"
Society of Petroleum Engineers of AIME, SPE 12206 (October 1983).
[a]February 1983 prices.

If properly and prudently done, heating an emulsion can greatly benefit water separation. However, if a satisfactory rate of water removal can be achieved at the minimum temperature delivered into a process, there may be no reason to suffer the economic penalties associated with adding heat.

1.4.10 Electrostatic Coalescers

Coalescing of the small water drops dispersed in the crude can be accomplished by subjecting the water-in-oil emulsion to a high-voltage electrical field. When a non-conductive liquid (oil) containing a dispersed conductive liquid (water) is subjected to an electrostatic field, the conductive particles or droplets are caused to combine by one of three physical phenomena:

- The droplets become polarized and tend to align themselves with the lines of electric force. In so doing, the positive and negative poles of the droplets are brought adjacent to each other. Electrical attraction brings the droplets together and causes them to coalesce.

- Droplets are attracted to an electrode due to an induced charge. In an AC field, due to inertia, small droplets vibrate over a larger distance than larger droplets promoting coalescence. In a DC field the droplets tend to collect on the electrodes, forming larger and larger drops until eventually they fall by gravity.
- The electric field tends to distort and thus weaken the film of the emulsifier surrounding the water droplets. Water droplets dispersed in oil and subjected to a sinusoidal alternating-current field will be elongated along the lines of force during the first half cycle. As they are relaxed during the low-voltage portion, the surface tension will pull the droplets back toward the spherical shape. The same effect is obtained in the next half of the alternating cycle. The weakened film is thus more easily broken when droplets collide, making coalescence more likely.

Whatever the actual mechanism, the electric field causes the droplets to move about rapidly in random directions, which greatly increases the chances of collision with another droplet. When droplets collide with the proper velocity, coalescence occurs.

The attraction between water droplets in an electric field is given by

$$F = \frac{K_s \varepsilon^2 (d_m)^6}{S^4} \text{ (with } S \geq d_m), \qquad (1.4)$$

where

F = attractive force between droplets,
K_S = constant for system,
ε = voltage gradient,
d_m = diameter of droplets,
S = distance between droplets.

This equation indicates that the greater the voltage gradient is, the greater the forces causing coalescence will be. However, experimental data show that at some gradient the water droplet can be pulled apart and a strong emulsion can be developed. For this reason, electrostatic treaters are normally equipped with a mechanism for adjusting the gradient in the field.

1.4.11 Water Droplet Size and Retention Time

The droplet diameter is the most important single parameter to control to aid in water settling since this term is squared in Stokes' law's settling equation. A small increase in diameter will create a much larger increase in settling velocity. Thus, in sizing treating equipment, it is necessary to predict a droplet diameter, which must be separated from the oil to meet a desired BS&W specification.

It would be extremely rare to have laboratory data of droplet coalescence for a given system. Qualitatively, we would expect droplet size to increase with retention time in the coalescing section and with heat input, which excites the system, leading to more collisions of small droplets. Droplet size could be expected to decrease with oil viscosity, which inhibits the movement of the particles and decreases the force of the collision. While it may be possible to predict the droplet size at the inlet to the treater, the shearing that occurs at the inlet nozzle and inlet diverter coupled with the coalescence that occurs at the oil–water interface cannot be determined. The treater represents a dynamic process, which cannot be adequately simulated by static laboratory tests.

The coalescence equation indicates that the oil–water interface zone is where nearly all of the coalescence occurs. Except for providing some minimal time for initial coalescence to occur, increasing retention time in a crude oil treating system may not be very cost-effective. Consequently, in most systems one would not expect retention time to have a significant impact on increasing the water droplet diameter.

The effect of temperature on water droplet size distribution is small. The temperature does, however, have a large effect on the oil viscosity. Since temperature and retention time have relatively small effects, an empirical relationship can be proposed relating droplet size distribution to oil viscosity alone. This relationship assumes sufficient retention time has been provided so initial coalescence can occur. Typically, retention times vary from 10–30 min, but values outside this range are also common.

If the water droplet size distribution in the oil to be treated were known, it would be possible to predict the size of droplets that must be removed to assure that a specific amount of water remains in the treated oil. Therefore, a relationship exists between the design BS&W content of the treated oil and the droplet size that must be removed for a set droplet size distribution. Since the droplet size distribution is a function of viscosity as stated above, the droplet size to be removed is related to both the required BS&W and the oil viscosity.

1.5 Treater Equipment Sizing

1.5.1 General Considerations

The major factors controlling the sizing of emulsion treating equipment are:

- heat input required,
- gravity separation considerations,
- settling equations,
- retention time equations, and
- water droplet size.

1.5.2 Heat Input Required

The heat input and thus the fuel required for treating depend on the temperature rise, amount of water in the oil, and flow rate. Heating water requires about twice as much energy as it does to heat oil. For this reason, it is beneficial to separate any free water from the emulsion to be treated with either a FWKO located upstream of the treater or an inlet FWKO system in the treater itself.

Assuming that the free water has been separated from the emulsion, the water remaining is less than 10% of the oil, and the treater is insulated to minimize heat losses, the required heat input can be determined from

Field units

$$q = 16Q_o\Delta T[0.5(SG)_o + 0.1], \tag{1.5a}$$

SI units

$$q = 1100Q_o\Delta T[0.5(SG)_o + 0.1], \tag{1.5b}$$

where

q = heat input, BTU/h (kW),
Q_o = oil flow rate, bopd (m^3/h),
ΔT = increase in temperature, °F (°C),
SG_o = specific gravity of oil relative to water.

1.5.3 Gravity Separation Considerations

Most oil-treating equipment relies on gravity to separate water droplets from the oil continuous phase, because water droplets are heavier than the volume of oil they displace. However, gravity is resisted by a drag force caused by the droplets' downward movement through the oil. When the two forces are equal, a constant velocity is reached, which can be computed from Stokes' law as (Stokes' law was derived in the *Gas–Liquid and Liquid–Liquid Separation* volume).

Field units

$$V_t = 1.78 \times 10^{-6}\frac{(\Delta SG)d_m^2}{\mu}, \tag{1.6a}$$

SI units

$$V_t = 5.44 \times 10^{-7}\frac{(\Delta SG)d_m^2}{\mu}, \tag{1.6b}$$

where

V_t = downward velocity of the water droplet relative to the oil
       continuous phase, ft/sec (m/sec),
d_m = diameter of the water droplet, μm,
ΔSG = difference in specific gravity between the oil and water,
μ   = dynamic viscosity of the oil continuous phase, centipoise
       (cp).

Several conclusions can be drawn from Stokes' law:

- The larger the size of a water droplet, the larger the square of its diameter and, thus, the greater its downward velocity will be. That is, the bigger the droplet size, the less time it takes for the droplet to settle to the bottom of the vessel and thus the easier it is to treat the oil.
- The greater the difference in density between the water droplet and the oil phase, the greater the downward velocity will be. That is, the lighter the crude, the easier it is to treat the oil. If the crude gravity is 10°API and the water is fresh, the settling velocity is zero, as there is no gravity difference.
- The higher the temperature, the lower the viscosity of the oil and, thus, the greater the downward velocity will be. That is, it is easier to treat the oil at high temperatures than at low temperatures (assuming a small effect on gravity difference due to increased temperature).

1.5.4 Settling Equations

The specific gravity difference between the dispersed water droplets and the oil should result in the water 'sinking' to the bottom of the treatment vessel.

Since the oil continuous phase is flowing vertically upward in both vertical and horizontal treaters previously described, the downward velocity of the water droplet must be sufficient to overcome the velocity of the oil traveling upward through the treater. By setting the oil velocity equal to the water settling velocity, the following general sizing equations may be derived:

Horizontal Vessels

Field units

$$dL_{\text{eff}} = 438 \frac{FQ_o\mu_o}{(\Delta SG)d_m^2},$$ (1.7a)

SI units

$$dL_{\text{eff}} = 5.0 \times 10^5 \frac{FQ_o\mu_o}{(\Delta SG)d_m^2}.$$ (1.7b)

If the treater has a spreader and a collector, then the spreader/collector short-circuiting factor is 1. If the treater lacks the spreader, collector, or both, then '*F*' should be some value greater than 1.

Vertical Vessels

Field units

$$d = 81.8 \left[\frac{FQ_o\mu_o}{(\Delta SG)d_m^2} \right]^{1/2}$$ (1.8a)

SI units

$$d = 25{,}230 \left[\frac{FQ_o\mu_o}{(\Delta SG)d_m^2} \right]^{1/2}$$ (1.8b)

First, solve Equations (1.8a) and (1.8b) for *d* assuming $F = 1$. If $d \le 48$ in. (1220 mm), this is the final answer. If $d > 48$ in. (1220 mm), then substitute $F = d/48$ ($F = d/1220$) into Equation (1.8).

Note that the height of the coalescing section for a vertical treater does not enter into the settling equation. The cross-sectional area of flow for the upward velocity of the oil is a function of the diameter of the vessel alone. This is a limiting factor in the capacity of vertical treaters. In a horizontal vessel, the cross-sectional area for flow for the upward velocity of the oil is a function of the diameter times the length of the coalescing section.

Gunbarrels

The equations for gunbarrels are similar to those for vertical treaters since the flow pattern and geometry are the same. However, gunbarrel tanks experience a great deal of short-circuiting due to uneven flow distribution. This is a result of the large tank diameter. The sizing equation for gunbarrels includes a short-circuiting factor '*F.*' This factor accounts for imperfect liquid distribution across the entire cross section of the treating vessel or tank and is a function of the flow conditions in the vessel. The larger the retention time, the larger the short-circuiting factor will be. It may be necessary to apply a short-circuiting factor for large vertical treaters as well.

Field units

$$d = 81.8 \left[\frac{FQ_o\mu_o}{(\Delta SG)d_m^2} \right]^{1/2},$$ (1.9a)

SI units

$$d = 25{,}230 \left[\frac{FQ_o\mu_o}{(\Delta SG)d_m^2} \right]^{1/2},$$ (1.9b)

where

d = minimum vessel internal diameter, in. (mm),
Q_o = oil flow rate, BOPD (m^3/h),
μ_o = oil viscosity, cp,
L_{eff} = length of coalescing section, ft (m),
ΔSG = difference in specific gravity between oil and water (relative to water),
d_m = diameter of water droplet, μm,
F = short-circuiting factor

Horizontal Flow Treaters
In horizontal flow settling, the water droplets settle perpendicular to the oil flow. By setting the oil retention time equal to the water settling time, the following equation may be used:

Field units

$$wL_{eff} = 800\left(\frac{Q_o\mu_o}{(\Delta SG)d_m^2}\right), \tag{1.10a}$$

SI units

$$wL_{eff} = 9.2 \times 10^8\left(\frac{Q_o\mu_o}{(\Delta SG)d_m^2}\right), \tag{1.10b}$$

where

L_{eff} = effective length for separation, ft (m),
w = effective width of flow channel, in. (mm).

The effective length is normally 75% of the separation length available. For example, in Figure 1.41 the effective length is 75% of the sum of L1 through L4. The effective width is approximately 80% of the actual channel width.

Note that the height of the flow channel drops out of Equations (1.10a) and (1.10b). This is because the oil retention time and the water settling time are both proportional to the height.

1.5.5 Retention Time Equations

The oil must be held at temperature for a specific period of time to enable de-emulsifying the water-in-oil emulsion. This information is best determined in the laboratory but, in the absence of such data, 20–30 min is a good starting point.

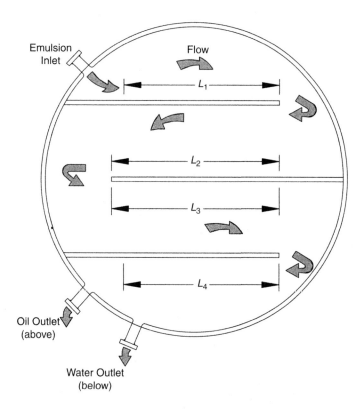

FIGURE 1.41. Plan view of a cylindrical treating tank using horizontal flow.

The retention time in the coalescing-settling section of a treater is the volume of the coalescing-settling section divided by the oil flow rate. The volume of the coalescing-settling section is a function of the vessel diameter and the height of the coalescing section.

Depending on the specific properties of the stream to be treated, the geometry required to provide a certain retention time may be larger or smaller than the geometry required to satisfy the settling equation. The geometry of the vessel is determined by the larger of the two criteria. The equations for retention time are as follows.

Horizontal Vessels

Field units

$$d^2 L_{\text{eff}} = \frac{Q_o(t_r)_o}{1.05} \qquad (1.11\text{a})$$

SI units

$$d^2 L_{\text{eff}} = \frac{Q_o(t_r)_o}{3.535 \times 10^{-5}}$$

(1.11b)

Vertical Vessels

Field units

$$d^2 h = \frac{(t_r)_o Q_o}{0.12}$$

(1.12a)

SI units

$$d^2 h = \frac{(t_r)_o Q_o}{4.713 \times 10^{-8}}$$

(1.12b)

Part of the overall vessel height is required to provide for water retention. The removal of oil from the water is not a primary concern. Equations can be derived for water retention similar to the equations for oil retention. Assuming that a short-circuiting factor is not critical, the height required for water retention can be derived.

The height of water required to provide a given retention time defines the distance between the down-comer exit and the water outlet. The height to the oil–water interface may be much greater due to the need to provide space for fire-tubes. The height of the coalescing section, and thus the overall height of the vessel, is most often determined by the need to maintain the oil at the oil–water interface above its bubble-point pressure. Thus, most vertical heater-treaters have much higher oil retention times than necessary for coalescence alone.

Gunbarrels

Field units

$$d^2 h = \frac{F(t_r)_o Q_o}{0.12},$$

(1.13a)

SI units

$$d^2 h = \frac{F(t_r)_o Q_o}{4.713 \times 10^{-8}},$$

(1.13b)

where

t_r = retention time, min,
Q_o = oil flow, BOPD (m^3/h),
h = height of the coalescing section, in. (mm),
F = short-circuiting factor

Horizontal Flow Treaters
The potential for short-circuiting in high tanks is great. Therefore, it is normally assumed that the height limit to consider in calculating retention time is 50% of the actual flow channel width. Providing higher flow channels neither increases the effective retention time nor increases the ability to separate water droplets from the oil.

To provide a specified oil retention time requires a certain volume based on flow rate as follows:

Field units

$$hwL_{\text{eff}} = 0.56(t_r)_o Q_o,$$ (1.14a)

SI units

$$hwL_{\text{eff}} = 1.67 \times 10^4 (t_r)_o Q_o,$$ (1.14b)

where h = effective height of the flow channel, in. (mm).

1.5.6 Water Droplet Size

In order to develop a treater design procedure, the water droplet size to be used in the settling equation to achieve a given outlet water cut must be determined. As previously mentioned, it would be extremely rare to have laboratory data of the droplet size distribution for a given emulsion as it enters the coalescing section of the treater. Qualitatively, we would expect the minimum droplet size that must be removed for a given water cut to (1) increase with retention time in the coalescing section; (2) increase with temperature, which tends to excite the system, leading to more collisions of small droplets; and (3) increase with oil viscosity, which tends to inhibit the formation of small droplets from shearing that occurs in the system.

We have seen that, after an initial period, increasing the retention time has a small impact on the rate of growth of particles. Thus, for practically sized treaters with retention times of 10–30 min, retention time would not be expected to be a determinant variable. Intuitively, one would expect viscosity to have a much greater effect on coalescence than temperature.

Assuming that the minimum required size of droplets that must be settled is a function only of oil viscosity, equations have been developed correlating this droplet size and oil viscosity (Thro and Arnold, 1994). The authors used data from three conventional treaters operating with 1% water cuts. Water droplet sizes were back-calculated using

Equation (1.7a) and (1.17b). The calculated droplet sizes were correlated with oil viscosity, and the following equations resulted:

$$d_{m1\%} = 200\mu^{0.25}, \ \mu_o < 80 \text{ cp}, \tag{1.15}$$

where

$d_{m1\%}$ = diameter of water droplet to be settled from the oil to achieve 1% water cut, μm,

μ = viscosity of the oil phase, cp.

Using the same procedure, the following correlation for droplet size was developed for electrostatic treaters:

$$d_{m1\%} = 170\mu^{0.4}, 3 \text{ cp} < \mu_o < 80 \text{ cp}. \tag{1.16}$$

For viscosities below 3 cp, Equations (1.8a) and (1.8b) should be used. The two equations intersect at 3 cp, and electrostatic treaters would not be expected to operate less efficiently in this range. Additionally, the data from which the electrostatic treater droplet size correlation was developed did not include oil viscosities less than 7 cp.

The same authors also investigated the effect of water cut on minimum droplet size. Data from both conventional and electrostatic treaters over a range of water cuts were used to back-calculate an imputed droplet size as a function of water cut, resulting in the following equation:

$$\frac{d_m}{d_{m1\%}} = W_c^{0.33}, \tag{1.17}$$

where

d_m = diameter of water droplet to be settled from the oil to achieve a given water cut (W_c), μm,

W_c = water cut, %.

As the volume of a sphere is proportional to the diameter cubed, Equations (1.11a) and (1.11b) indicate that the water cut is proportional to the droplet diameter cubed.

It must be stressed that the above equations should be used only in the absence of other data and experience. These proposed relationships are based only on limited experimental data.

An approximate sizing relationship, derived from Equations (1.15) and (1.16), are given in Figures 1.42 and 1.43 in terms of the flow rate of emulsion (given in BPD) flowing vertically through a horizontal cross-sectional area of one square foot. For a horizontal treater with vertical flow through the coalescing section, the flow area can be approximated as the diameter of the vessel times the length of the coalescing section.

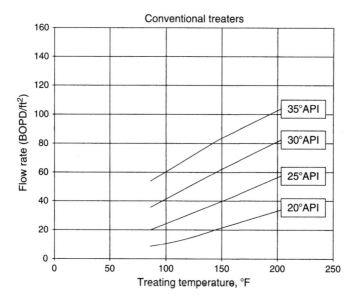

FIGURE 1.42. Flow rate versus treating temperature for conventional treaters.

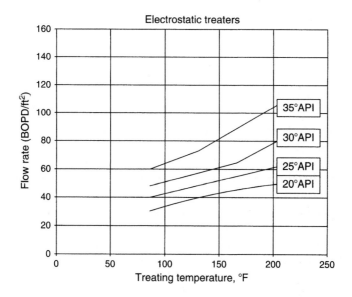

FIGURE 1.43. Flow rate versus treating temperature for electrostatic treaters.

1.6 Design Procedure

In specifying the size of a treater, it is necessary to determine the diameter (d), length or height of the coalescing section (L_{eff} or h), and treating temperature or fire-tube rating. As we have seen, these variables are interdependent, and it is not possible to arrive at a unique solution for each. The design engineer must trade the cost of increased geometry against the savings from reducing the treating temperature.

The equations previously presented provide tools for arriving at this trade-off. However, because of the empirical nature of some of the underlying assumptions, engineering judgment must be utilized in selecting the size of treater to use.

1.6.1 General Design Procedure

1. Choose a treating temperature.
2. Determine the heat input required from Equations (1.5a) and (1.5b).
3. Determine oil viscosity at treating temperature. In the absence of laboratory data, Figure 1.33 provides a correlation that can be used to estimate crude viscosity given its gravity and temperature.
4. Select a type of treater, and size the treater using the appropriate design procedure below.
5. Choose the design minimum droplet size that must be separated from experimentel data, analogy to other treaters in services or Eqs. 1.14, 1.15 and 1.16.
6. Repeat the above procedure for different treating temperatures.

1.6.2 Design Procedure for Vertical Heater-Treaters and Gunbarrels (Wash Tanks with Internal/External Gas Boot)

1. Calculate the minimum treater diameter using Equations (1.8a)–(1.9b).
2. For various diameters greater than the minimum, calculate the height required in the coalescing-settling section using Equations (1.12a)–(1.13b).
3. Calculate the height required to keep the oil above its bubble-point pressure if the emulsion is heated after the gas has been separated from it. For most standard applications, a 20- or 27-ft (6.1 or 8.2 m) height is acceptable.
4. Select a standard size treater from vendor literature that meets the above requirements. *Note:* Standard sizes are presented in Figures 1.44 and 1.45.

Typical Vertical Treater Dimensions and Pressures

Outside Diameter ft	Shell Length Head Seam to Head Seam to ft = 6 in.	Minimum Design Pressure psig
3	10.12 or 15	50
4	10.12.20 or 27 1/2	50
6	12.20. or 27 1/2	50
8	20 or 27 1/2	40
10	20 or 27 1/2	40

Typical Horizontal Treater Dimensions and Pressures

Outside Diameter ft	Shell Length Head Seam to Head Seam to ft = 6 in.	Minimum Design Pressure psig
3	10.12 or 15	50
4	10.12 or 15	50
6	12.15 or 20	50
8	15.20.25 or 30	50
10	20.30.40.50 or 60	50
12	30.40.50 or 60	50

TYPICAL FIREBOX RATINGS

Outside Diameter ft	Vertical		Horizontal	
	Minimum Area sq ft	Heat Duty BTU/hr	Minimum area sq ft	Heat Duty BTU/hr
3	10	100,000	15	150,000
4	25	250,000	25	250,000
6	50	500,000	50	500,000
8	100	1,000,000	75	750,000
10	125	1,250,000	200	2,000,000
12	-	-	320	3,200,000

These are suggested minimum requirements. Heat requirements of the process and fluid characteristics should be considered in sizing the firebox.

FIGURE 1.44. Standard dimensions, pressure ratings, and fire-box ratings for vertical and horizontal heater-treaters.

VFH OIL PROCESSING UNIT CAPACITIES

Shell size (Length× Diameter)	Fire-tube capacity (BTU/Hr)	Fire-tubes (Number and O.D)	Oil		Free Water (Barrels per day)	Gas (MM sct/d)
			(Barrel per hour)	(Barrel per day)		
STANDARD MODEL - 8° to 42°API GRAVITY OIL						
6'×15'	550,000	1–18"	15–75	360–1800	1500	0.5–1
6'×20'	1,000,000	1–18"	35–120	800–2800	2,200	0.5–1
6'×30'	1,500,000	1–18"	70–180	1500–4000	3,500	0.5–1
8'×20'	1,375,000	1–24"	40–130	1000–3200	3,500	1.5–2
8'×25'	1,750,000	1–24"	60–160	1400–3800	4,500	1.5–2
8'×25'	2,500,000	2–18"	60–160	1400–3800	4,500	1.5–2
9'×30'	2,000,000	1–24"	75–200	1800–4800	5,500	1.5–2
9'×30'	3,000,000	2–18"	75–200	180–4800	5,500	1.5–2
8'×35'	2,500,000	1–24"	85–220	2000–5300	6,500	1.5–2
8'×35'	3,750,000	2–18"	85–220	2000–5300	6,500	1.5–2
10'×25'	2,250,000	2–18"	100–250	2400–6000	5,000	2–3
10'×25'	3,000,000	2–24"	100–250	2400–6000	5,000	2–3
10'×30'	2,500,000	2–18"	120–320	2900–7700	6,000	2–3
10'×30'	3,500,000	2–24"	120–320	2900–7700	6,000	2–3
10'×35'	2,750,000	2–18"	150–400	3600–9600	6,500	2–3
10'×35'	3,750,000	2–24"	150–400	3600–9600	6,500	2–3
10'×40'	2,750,000	2–18"	190–500	4500–12000	7,000	2–3
10'×40'	3,750,000	2–24"	190–500	4500–12000	7,000	2–3
12'×45'	5,000,000	2–24"	190–500	4500–12000	8,500	2–3
12'×75'	6,000,000	2–24"	200–520	4800–12500	10,500	2–3
12'×45'	7,500,000	3–24"				
12'×50'	9,400,000	3–24"				
12'×55'	9,400,000	3–24"				
12'×60'	10,500,000	3–24" PLEASE CONSULT C-E NATCO ENGINEERING				
12'×65'	10,500,000	3–24" FOR CAPACITIES OF 12' DIAMETER UNITS				
12'×70'	13,000,000	3–24"				
12'×75'	13,000,000	3–24"				
ROCKY MOUNTAIN MODEL - 28°C to 42°API GRAVITY OIL 50 PSI WORKING PRESSURES						
6'×15'	650,000	1–18"	12–60	300–1400	1000	0.5–1
6'×20'	850,000	1–18"	30–90	700–2200	1500	0.5–1
6'×25'	1,250,000	1–18"	36–100	900–2500	1700	0.5–1
25 AND HIGHER PSI WORKING PRESSURES AVAILABLE						
8'×15'	750,000	1–24"	20–75	500–1800	1600	1.5–2
8'×15'	1,000,000	2–18"	30–90	700–2100	1600	1.5–2
8'×20'	1,375,000	1–24"	40–100	1000–2400	2300	1.5–2
8'×20'	1,500,000	2–18"	60–130	1500–3200	2000	1.5–2
8'×25'	1,250,000	1–24"	75–150	1800–3600	3000	1.5–2
8'×25'	1,750,000	2–18"	75–150	1800–3600	3000	1.5–2
8'×30'	1,500,000	1–24"	80–170	2000–4200	3600	1.5–2
8'×30'	1,750,000	2–18"	80–170	2000–4200	3600	1.5–2
8'×35'	1,650,000	1–24"	90–200	2200–4800	4200	1.5–2

FIGURE 1.45. Typical vendor supplied vertical heater-treater capacity table.

1.6.3 Design Procedure for Horizontal Heater-Treaters

1. For various standard diameters, develop a table of effective lengths versus standard diameters, using Equations (1.10a) and (1.10b) for settling.
2. For the same diameters used in step 1, calculate the effective lengths required using Equations (1.11a) and (1.11b) for retention time.
3. Select a treater that satisfies the larger effective length requirements for the selected diameter.

1.6.4 Design Procedure for Horizontal-Flow Treaters

1. Calculate several combinations of w and L_{eff} using Equations (1.10a) and (1.10b) for settling.
2. For each combination of w and L_{eff} used in step 1, calculate the h required for the specified retention time using Equations (1.14a) and (1.14b).
3. Select a combination of w and L_{eff} for which the calculated h is less than one half the flow width.

The above procedure allows the production facility engineer to choose the major sizing parameters of heater-treaters when little or no laboratory data are available. This procedure does not give the overall dimensions of the treater, which must include inlet gas separation and FWKO sections. However, it does provide a method for specifying a fire-tube capacity and a minimum size for the coalescing section (where the treating actually occurs) and provides the engineer with the tools necessary to evaluate specific vendor proposals.

Figure 1.44 provides standard dimensions, pressure ratings, and fire-box ratings for vertical and horizontal heater-treaters. Figure 1.45 is a typical horizontal heater-treater table supplied by an equipment manufacturer. Figure 1.46 is a typical vertical heater-treater capacity table supplied by an equipment manufacturer.

1.7 Practical Considerations

Successful treatment of emulsions, depending on specific emulsion characteristics, can be treated by low temperature with or without adding chemicals, or chemicals with or without heat. Some fields having high water cut (e.g., 95%) can be treated successfully without heat or chemicals, but require extremely long retention times. It is better to use chemicals instead of heat from the standpoints of installation, maintenance, and operating costs. The following discussion provides some general guidelines to help one select the right oil treating equipment configuration for a specific application.

Shell size (Length × Diameter)	Fire-tube capacity (Bt/Hr)	Fire-tubes (Number and O.D)	Oil				Free Water (barrels per Day)	Gas (MM sctd)
			Bbls/Hr		Bbls/Day			
			AC	AC/DC	AC	AC/DC		
6'×15'	550,000	1.18"	20–100	24–120	480–2400	576–2880	500–1500	0.5–1
6'×20'	1,000,000	1.18"	20–100	24–120	480–2400	576–2880	500–1500	0.5–1
8'×15'	750,000	1.24"	50–180	60–261	1200–4320	1400–5184	600–1800	1.5–2
8'×15'	1,100,000	2.18"	50–180	60–261	1200–4320	1400–5184	600–1800	1.5–2
8'×20'	1,300,000	1.21"	100–230	120–276	2400–5520	2880–6624	800–2400	1.5–2
8'×20'	2,000,000	2.18"	100–230	120–276	2400–5520	2880–6624	800–2400	1.5–2
8'×25'	1,500,000	1.24"	125–250	150–300	3000–600	3600–7200	800–2400	1.5–2
8'×25'	2,250,000	2.18"	125–250	150–300	3000–600	3000–6000	800–2400	1.5–2
10'×20'	2,000,000	2.18"	140–280	168–336	3360–6720	4032–8064	1000–3000	2–3
10'×20'	2,500,000	2.24"	140–280	168–336	3360–6720	4032–8064	1000–3000	2–3
10'×20'	3,000,000	3.18"	140–280	168–336	3360–6720	4032–8064	1000–3000	2–3
10'×25'	2,000,000	2.18"	175–430	210–516	4200–10320	5040–12384	1000–3000	2–3
10'×25'	2,500,000	2.24"	175–430	210–516	4200–10320	5040–12384	1000–3000	2–3
10'×25'	3,000,000	3.18"	175–430	210–516	4200–10320	5040–12384	1000–3000	2–3
10'×30'	2,000,000	2.18"	200–580	240–696	4800–13920	5760–16704	1000–3000	2–3
10'×30'	2,500,000	2.24"	200–580	240–696	4800–13920	5760–16704	1000–3000	2 3
10'×30'	3,000,000	3.18"	200–580	240–696	4800–13920	5760–16704	1000–3000	2–3
10'×35'	3,000,000	2.24"	200–580	240–696	4800–13920	5760–16704	1500–4500	2–3
10'×35'	3,750,000	3.18"	200–580	240–696	4800–13920	5760–16704	1500–4500	2–3
10'×40'	3,750,000	2.24"	350–730	420–876	8400–17520	10080–21024	2000–6000	3–5
10'×45'	5,000,000	2.24"	350–730	420–876	8400–17520	10080–21024	2500–7500	3–5
10'×50'	6,000,000	2.24"	350–730	420–876	8400–17520	10080–21024	3000–9000	3–5

FIGURE 1.46. Typical vendor-supplied horizontal electrostatic heater-treater capacity table.

1.7.1 Gunbarrels with Internal/External Gas Boot

Gunbarrels (wash tank with internal/external gas boot) should be considered when isolated, high saltwater percentage production is indicated, provided retention time requirements do not make gunbarrel sizing impractical. When used without heat, the vessel should provide ample settling time, for example, 12–24 h. Sufficient retention time allows some storage of basic sediment during cold weather when chemical efficiency declines. The basic settlement is cleaned from the tank during warm weather and by periodically rolling (circulating) the gunbarrel.

1.7.2 Heater-Treaters

A heater-treater should be considered in fields requiring heat to break the emulsion. Good practice is to install a slightly larger (+10%) heater-treater than is necessary. This allows extra capacity for unforeseeable production increases (normally water), reduction in the amounts of treating chemical used, and startup of a cold unit. A reduction in chemical cost can easily pay for the additional cost of a larger treater in a few years. Depending on the characteristics of the oil and the efficiency of the chemical, retention times range between 10 and 60 min.

1.7.3 Electrostatic Heater-Treaters

An electrostatic heater-treater should be considered in fields with maximum salt content specifications imposed [10–30 lb per thousand barrels (PTB)], any time the BS&W must be reduced below 0.5%, and offshore facilities where space and/or heat is limited.

Other configuration considerations that the designer may be required to evaluate are FWKO instead of a gunbarrel and using an electrostatic heater-treater instead of a heater-treater. Applying the basic principles presented in this section coupled with sound engineering judgment will allow the designer to select the most appropriate selection.

1.8 Oil Desalting Systems

1.8.1 Introduction

The process of removing water-soluble salts from an oil stream is called oil desalting. Nearly all crude oil is produced with some entrained water, which normally contains dissolved salts, principally chlorides of sodium, calcium, and magnesium. The majority of the produced salt water is removed in the separation and treating process. However, a small amount of entrained water remains in the crude oil. The crude oil is sent to the refinery where it is heated as part of the various refinery processes. The entrained water is driven off as steam. However, the salts in the water does not leave with the steam but crystallizes and remains suspended in the oil or may deposit as scale within heat exchange equipment. In addition, entrained salt crystals will usually deactivate catalyst beds and plug downstream processing equipment.

Due to these problems, refineries usually reduce crude oil salt contents to very low levels prior to processing. Refineries usually achieve the needed salt content by specifying in purchase contracts a maximum salt content, as well as maximum water content. A common salt specification would be 10 pounds per thousand barrels (PTB), 10 PTB (0.003 kg/m^3). To satisfy the refinery specification, upstream production facilities may be required to perform some oil desalting.

This part of the chapter describes the methods and equipment commonly used to desalt crude oil.

1.8.2 Equipment Description

Desalters

Since the salt content is directly related to the amount of residual water, the best desalters remove as much water as possible. Any device that removes water from oil can be used as a desalter. However, the majority of desalters employed are horizontal electrostatic treaters. These treaters will produce the lowest residual water level of all treaters. Figure 1.28 illustrates a horizontal electrostatic treater of the type typically used in desalting operations. Because very low water contents are required, the crude is usually pumped through the desalter at pressures above its bubble-point. In addition, the temperature of the crude to be desalted is determined by upstream heat exchangers. Thus, there is need for an inlet degassing and heating section as shown in the typical oil field horizontal electrostatic treater discussed earlier.

1.8.3 Mixing Equipment

Globe Valves

A manual globe throttling valve is one of the simplest methods to promote the mixing of dilution water and salt water. The pressure drop resulting from forcing the oil and water through a throttling valve may be used to shear the water and mix the droplets in the oil. The major disadvantage of any manual valve is its inability to automatically adjust for changes in oil flow rate. As the flow rate varies, the pressure drop—and thus the mixing efficiency—varies. Therefore, if the oil flow rate increases significantly, the pressure drop may increase to the point that the resulting mixed emulsion is almost impossible to treat.

It is possible to automate the globe valve to avoid "over mixing." A differential pressure controller is used to control the pressure drop through the globe valve. This system automatically adjusts for changing flow rates and maintains a set pressure drop. Since this system's set point can be adjusted in the field, it allows an operator to optimize its performance.

The conventional single-ported and balanced double-ported globe valves are commonly used and yield good results. The valve body should be line size. If a single-port valve is used, some form of premixing should be provided prior to the valve to ensure an even distribution of the water droplets to control the droplet size distribution. Double-port globe valves yield lower pressure drops than single-port valves, and may eliminate the need for premixing.

The pressure drop through the mixing valve varies from approximately 10–50 psi (70–340 kPa). The required pressure drop through the mixing valve varies from 10–50 psi (70–340 kPa). The required pressure drop can be decreased if a premixing device is installed upstream of the mixing valve. The reason for this is that the premixing device distributes the water in the oil, and the valve will be required only to shear the droplets. If the mixing must both shear and distribute the water, higher pressure drops are necessary.

Spray Nozzles

Upstream premixing is commonly performed with either spray nozzles or static mixers.

As shown in Figure 1.47, one common method of premixing the water and oil involves using a system of spray nozzles. Water is pumped through the nozzles and then distributed throughout the oil stream. These systems are effective and are usually less expensive than static mixers.

Static Mixers

Static mixers use pieces of corrugated plate, as shown in Figure 1.48. These mixers typically divide into many parallel paths which divide and recombine as the flow passes through the mixer. The alternate layers of corrugations are perpendicular to each other so that the fluid must pass through a series of relatively small openings. This mixer shears the water droplets to a much smaller size than the old mixers. These mixers produce a narrow range of droplet sizes. This is a result of two opposing phenomena. Large droplets are sheared by the mixing action in the small openings, while at the same time these mixers provide large surface areas where small droplets may collect and coalesce. Theoretically, the coalescing ability improves the performance of the

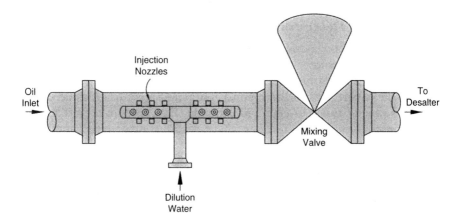

FIGURE 1.47. Schematic of a spray nozzle system for premixing water and oil.

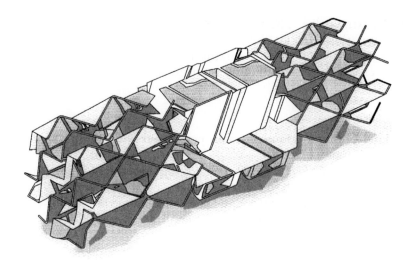

FIGURE 1.48. New static mixer.

dehydration equipment due to the reduction in the number of very small droplets which makes dehydration easier and decreases the chances of creating a stable, untreatable emulsion during the mixing process.

Static mixers are sized to provide an average droplet size using empirical equations based on test data. The average droplet size for desalting should be roughly between 250 and 500 μm. The average droplet size is a function of the oil flow rate. The primary disadvantage of static mixers is that they may not be adjusted as the flow varies. Therefore, if the oil flow will vary over a range of 3–1, or more, static mixers should not be used as the only mixing device.

1.8.4 Process Description

Most of the salt contained in crude oil is dissolved in the small water droplets. Since water is the salt carrier, removing the water will remove the salt from the crude. The salt content of the water is expressed as parts per million (ppm) equivalent sodium chloride. Salinity may range from 0 to over 150,000 ppm. Desalting is required when the amount of salt contained in the entrained water after treating is higher than some specified amount.

For example, assume a heater-treater is used for dehydration and it yields oil that is 0.5% water, each thousand barrels of dehydrated oil includes 5 bbl of water. If we next assume the water has a low salt content, say 10,000 ppm NaCl, then each barrel of water would contain approximately 3.5 lb of salt. With 5 bbl of water per thousand barrels of oil, the oil would then contain approximately 17.5 PTB. If the purchase agreement specified 10 PTB or less, some desalting, or a more efficient dehydrator, would be required.

In this example, an electrostatic treater might be all that is required to achieve an oil outlet that contains less than 0.3% water. This example assumed a low salt content. If the water had a high salt content, say 200,000 ppm NaCl, there would be approximately 70 lb of salt per barrel of water (lb/bbl). In this case, even dehydrating to 0.1% leaves 70 PTB. To reach the required 10 PTB, desalting would be required.

The desalting process involves two steps. The first step is to mix fresh water with entrained produced water. This will lower the produced water salinity by diluting the salt. The second step is dehydration, which is the removal of water from the crude. This dilution and dehydration produces a lower salinity in the residual water in the crude oil. The dilution water in desalting does not have to be fresh. Any water with a lower salt content than the produced water can be used.

Single-Stage Desalting

Figure 1.49 is a schematic of a single-stage desalting system. In this system, the dilution water is injected into the oil stream and then mixed. The oil then enters the desalter where the water is removed. To reduce dilution water requirements, the crude oil may be dehydrated prior to the desalting process. This removes the bulk of the produced water prior to desalting.

Two-Stage Desalting

Figure 1.50 is a schematic of a two-stage desalting system with dilution water recycling capability. This system is similar to the dehydrator and desalter system described in the previous section. The only difference is that the water removed in the second stage is pumped back to the first stage. The addition of this recycle provides for some

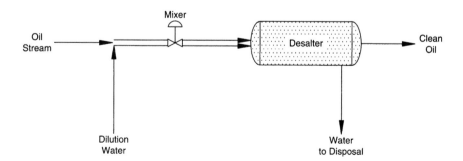

FIGURE 1.49. Schematic of a single-stage desalting system.

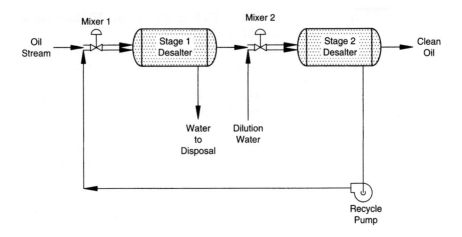

FIGURE 1.50. Schematic of a two-stage desalting system.

dilution of the salt water prior to the first stage. This further reduces the dilution water requirement compared to a single-stage dehydrator and desalter system.

If further desalting is needed, it is possible to add more stages in a similar manner.

References

Arnold, K. E. and Stewart, M. I. *Surface Production Operations-Design of Oil Handling Facilities*, 3rd Edition, Elsevier, 2008.

Thro, M. E. and Arnold, K. E. *Water Droplet Size Determination for Improved Oil Treater Sizing*, Society of Petroleum Engineers 69th Annual Technical Conference and Exhibition, New Orleans, LA, 1994.

CHAPTER 2

Crude Stabilization

2.1 Introduction

Crude oil or condensate stabilization is the removal of light components from a hydrocarbon liquid to lower its vapor pressure to a desired level. Stabilization may be used to meet a required pipeline sales contract specification or to minimize the vaporization of the hydrocarbon liquid stored in an atmospheric stock-tank. The stabilization process also results in reducing the amount of intermediate hydrocarbon components (propane and butane) that flash to the vapor state, increasing the liquid volume. Thus, it results in both increasing the liquid sales and decreasing the vapor pressure of the liquid.

Various methods are used to remove the light components from hydrocarbon liquid, with the most common being stage separation before the oil enters a stock-tank or pipeline. Although separation followed by weathering in a stock-tank is not the most efficient method of stabilization, it is often the most economical method. A stabilizer can achieve a stable specification product with a higher recovery, but with correspondingly higher capital investment and operating costs. Additional space is also required for a stabilizer. While this may not be a factor for onshore applications, it is often a major consideration for an offshore installation.

The purpose of this chapter is to describe the various processes used to stabilize a crude oil or condensate stream.

2.2 Theory of Crude Stabilization

2.2.1 General

The hydrocarbon fluid produced from wells normally flows to a separator for removal of the hydrocarbon gas. The hydrocarbon liquid or crude oil outflow from the separator will then go through additional stages of treatment before reaching the sales point. One of the objectives in the further treatment of the hydrocarbon liquid is stabilization of the crude. One or more of the following methods of crude stabilization are normally used:

- weathering in a stock-tank,
- multi-stage separation,
- crude oil treater after separation,
- crude oil stabilizer.

The method selected for crude stabilization depends upon a number of factors, principally economics and sales specifications. A crude oil stabilizer will almost always result in increased liquid recovery, but the capital investment and operating costs are correspondingly higher. Some factors that favor the installation of a crude stabilizer are:

- An oil sales contract requiring a low crude vapor pressure that cannot easily be obtained by stage separation.
- A sour crude with an oil sales contract limiting H_2S content to less than 60 ppm.
- Condensate production with 50 °API or higher and flow rates in excess of 33 m^3/h (5000 BPD).
- Fields where heavy components are vaporized into the gas phase but there is no sales contract for the gas.

2.2.2 Phase-Equilibrium Calculations

To understand and evaluate methods of crude stabilization, one must be familiar with phase equilibrium. Almost every operation in the production of hydrocarbons involves some form of equilibrium between the vapor and liquid phases of multi-component hydrocarbon systems. The distribution of individual components between phases has been correlated in terms of equilibrium ratios, or K values, which are functions of the temperature, pressure, and composition of the system.

   Before discussing equilibrium calculations, one should understand the nature of the equilibrium relationships in multi-component hydrocarbon systems and the regions in which each calculation is applicable. A typical pressure–temperature diagram is shown in Figure 2.1. A specific diagram of this type could be drawn from any system of fixed composition. The actual pressure and temperature coordinates will be different for various compositions.

   The equilibrium relationships apply only at pressure and temperature combinations in the two-phase region, or the area between the

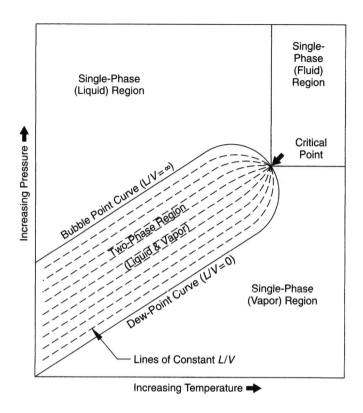

FIGURE 2.1. Phase diagram for a multi-component hydrocarbon system.

dew-point and bubble-point curves shown in Figure 2.1. The dew-point curve represents that point at which the first drop of liquid forms from a vapor phase system of fixed composition. Conversely, the bubble-point curve represents the point at which the first bubble of vapor forms from a liquid phase system. In the high-temperature/low-pressure region, the hydrocarbon mixture will be in a single vapor phase; in the high-pressure/low-temperature region, the hydrocarbon mixture will be in the single liquid phase. In these single-phase regions, the equilibrium relationships are not applicable.

Also shown in Figure 2.1 is a critical point of the system, where the dew-point and bubble-point curves converge. This is known as the convergence pressure for the system. In the region above the critical temperature and pressure, the hydrocarbon mixture exists as a single phase in which the vapor and liquid phases are indistinguishable. Refer to *Gas–Liquid and Liquid–Liquid Separation* volume for a detailed discussion on phase equilibrium.

2.2.3 Flash Calculations

Equilibrium separation involves the two-phase region between the two curves. Equilibrium calculations, often referred to as "flash calculations," are based upon a combination of the vapor—liquid equilibrium relationship and material balance equations.

The amount of hydrocarbon vapor and liquid, at any point in the process, is determined by a flash calculation. For a given pressure and temperature, each component of the hydrocarbon mixture will be in equilibrium. The mole fraction of the component in the gas phase will depend not only on pressure and temperature but also on the partial pressure of that component. Therefore, the amount of gas depends upon the total composition of the fluid, since the mole fraction of any one component in the gas phase is a function of the mole fraction of every other component in this phase.

The distribution of individual components between phases has been correlated in terms of equilibrium ratios, or K values, which are functions of the temperature, pressure, and composition of the system. This distribution is defined as:

$$K_N = \frac{V_N/V}{L_N/L} = \frac{y_N}{x_N},\tag{2.1}$$

where

K_N = constant for component N at a given temperature and pressure,

V_N = moles of component N in the vapor phase,

V = total moles in the vapor phase,

L_N = moles of component N in the liquid phase,

L = total moles in the liquid phase,

y_N = vapor mole fraction of component N,

x_N = liquid mole fraction of component N.

Much effort has gone into providing the industry with the best available information on equilibrium ratios. One source, the Gas Processors Suppliers Association (GPSA) (11th edition, Section 25), presents graphs of K values for the important components in a hydrocarbon mixture. The K values are for a specific "convergence" pressure, which is a function of the composition of the vapor and liquid phases. The convergence pressure is not a function of the pressure of the system, but if the liquid has been exposed to low pressure, allowing the light ends to boil off, then the convergence pressure will be lower.

In most oil-field applications, the convergence pressure will be between 2000 and 3000 psia (14,000 and 21,000 kPa), except for compositions that have been exposed to very low pressures, where convergence pressures between 500 and 1500 psia (3400 and 10,000 kPa) are possible. If the operating pressure is much less than the convergence pressure, the equilibrium constant is not greatly affected by the choice of convergence pressure. However, the choice of the convergence pressure becomes more critical the closer the operating pressure is to the convergence pressure. Therefore, using a convergence pressure of 2000 psia (14,000 kPa) is a good first approximation for most flash calculations where the operating pressure is less than 800 psi (5500 kPa). For operating pressures of 800–1500 psia (5500–10,000 kPa), a 3000-psia (21,000 kPa) convergence pressure should be used. Where greater precision is required, the convergence pressure should be calculated using the procedure in the GPSA's *Engineering Data Book*.

If K_N for each component and the ratio of total moles of vapor to total moles of liquid (V/L) are known, then the moles of component N in the vapor phase (V_N) and the moles in the liquid phase (L_N) can be calculated from:

$$V_N = \frac{K_N F_N}{\dfrac{1}{(V/L)} + K_N}, \tag{2.2}$$

$$L_N = \frac{F_N}{K_N(V/L) + 1}, \tag{2.3}$$

where F_N is the total moles of component N in the fluid.

To solve either Equation (2.2) or (2.3), it is first necessary to know the quantity V/L, but since both V and L are determined by summing V_N and L_N, it is necessary to use an iterative solution. This is done by first estimating V/L, calculating V_N and L_N for each component, summing up to obtain the total moles of gas (V) and liquid (L), and then comparing the calculated value (V/L) to the assumed value. In performing this procedure, it is helpful to use the following relationship:

$$L = \frac{F}{1 + (V/L)} \tag{2.4}$$

Once a value of V/L has been assumed, it is easy to calculate the corresponding value of L_N.

2.2.4 Dew-Point

The bottom line of Figure 2.1 represents the dew-point curve of the multi-component hydrocarbon system. The dew-point is the point at which the first drop of liquid forms from a vapor phase system of fixed composition. Mathematically, it may be expressed as the point at which

L/V approaches 0 in going from the two-phase region into the vapor phase region. As L/V approaches 0, the number of moles of component N in the vapor approaches the total number of moles of component N in the feed, and the total number of moles in the vapor phase approaches the total number of moles in the feed. The equilibrium relationship at the dew-point is:

$$\sum \frac{F_N}{K_N} = \sum F_N. \tag{2.5}$$

Any combination of pressure and temperature that produces K_N values for the given system which satisfy this equation is a dew-point condition. The vapor stream from an equilibrium separation is at its dew-point when separated.

2.2.5 Bubble-Point

The top line of Figure 2.1 represents the bubble-point curve, which is the point at which the first bubble of vapor forms from a liquid-phase system. Mathematically, it may be expressed as the point at which L/V approaches infinity in going from the two-phase region into the liquid-phase region. As L/V approaches infinity, the number of moles of liquid approaches the number of moles in the feed. At the bubble-point:

$$\sum K_N F_N = \sum F_N. \tag{2.6}$$

Any combination of pressure and temperature that produces K_N values for the given system which satisfy Equation (2.6) is a bubble-point condition. The liquid stream from any equilibrium separation is at its bubble-point when separated, just as the vapor stream is at its dew-point. Any flash separation divides the hydrocarbon system into two streams: a bubble-point stream and a dew-point stream.

The bubble-point pressure of a hydrocarbon liquid at a given temperature is the vapor pressure of that liquid at the same temperature. If the composition and temperature of a hydrocarbon liquid are known, its vapor pressure can be determined by calculating the bubble-point pressure.

It can be seen from the above that if the vapor pressure of a hydrocarbon liquid at ambient temperature is less than atmospheric pressure, the liquid can be stored in an atmospheric tank with minimal loss or shrinkage. On the other hand, if the vapor pressure of a hydrocarbon liquid is higher than atmospheric pressure at ambient temperature, and the liquid is stored in an atmospheric tank, some of the product will flash to atmosphere, resulting in loss of product.

2.2.6 Multi-Stage Separation

The most common method of separating oil and gas involves a *series of separators* and is termed "multi-stage separation." A system of this type typically requires from two to four separation steps, each occurring in a separator vessel. In addition to the separation steps, weathering may occur in the stock-tank. Figure 2.2 is a schematic of a separation process with three stages of separators followed by a final separation in the stock tank.

All of the principles of equilibrium separation apply to multi-stage separation. The liquid leaving each separator is at its bubble-point, and the liquid's vapor pressure is the same as the separator operating pressure at the separation temperature. As the liquid is reduced in pressure, hydrocarbon vapors are formed. The amounts and compositions of the hydrocarbon liquids and vapors can be determined by running flash calculations at the various conditions.

2.2.7 Initial Separator Pressure

Because of the multi-component nature of the produced fluid, a higher initial separation pressure will cause more liquid to be obtained in the separator. This liquid contains some light components, which vaporize in the stock-tank downstream of the separator. If the pressure for the initial separation is too high, too many light components will stay

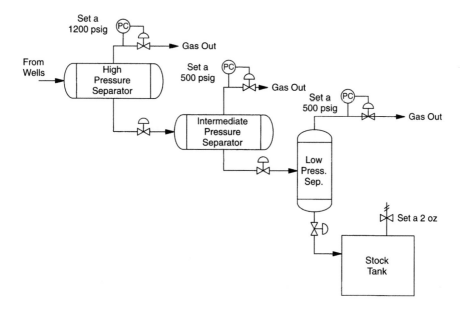

FIGURE 2.2. Schematic of a three-stage separation system.

in the liquid phase at the separator and be lost to the gas phase at the tank. If the pressure is too low, too many of the light and intermediate components will be lost to the gas phase in the separator, resulting in less liquid going to the stock-tank.

This phenomenon can be calculated using flash equilibrium techniques. It is important that we try to understand this phenomenon qualitatively. The tendency of any one component in the process stream to flash to the vapor phase depends on its partial pressure. The partial pressure of a component in a vessel is defined as the number of molecules of that component in the vapor space divided by the total number of molecules of all components in the vapor space times the pressure in the vessel. Since the partial pressure of a component is a function of the system's pressure, an increase in the system's pressure will increase the component's partial pressure. The increase in partial pressure reduces the component's equilibrium constant, and the molecules of that component tend toward the liquid phase. As the separator pressure is increased, the liquid flow rate out of the separator increases.

The problem with this process is that many of these molecules are the lighter hydrocarbons (methane, ethane, and propane), which have a strong tendency to flash to the gas state at stock-tank conditions (atmospheric pressure). In the stock-tank, the presence of a large number of light hydrocarbon molecules results in an increased tendency of intermediate-range hydrocarbons (butanes, pentane, and heptanes) to flash. Thus, by keeping the lighter molecules in the feed to the stock-tank, we capture a small amount of them as liquids, but we lose to the gas phase many more of the intermediate-range molecules. This is why, beyond some optimum point, there is actually a decrease in stock-tank liquids if the separator's operating pressure increases.

2.2.8 Stage Separation

In a simple single-stage process, the fluids are flashed in an initial separator, and then the liquids from that separator are flashed again at the stock-tank. By custom, the stock-tank is not normally considered a discrete stage of separation, though in reality it is.

Figure 2.2 shows a three-stage separation process. The liquid is first flashed at an initial pressure and then twice more at successively lower pressures before entering the stock-tank.

Because of the produced fluid's multi-component nature, flash calculations show that additional stages of separation result in more light components being stabilized into the liquid phase. This can be understood qualitatively by realizing that in a stage separation process the light hydrocarbon molecules that flash are removed at relatively high pressure, keeping the partial pressure of the intermediate hydrocarbons lower at each stage. As the number of stages approaches

TABLE 2.1
Effect of separation pressure for a rich condensate stream (Field Units)

Case	Separation Stage Pressures (Psia)	Liquid Produced (BOPD)	Compressor HP Required (HP)
I	1215; 65	8400	861
II	1215; 515; 65	8496	497
III	1215; 515; 190; 65	8530	399

TABLE 2.2
Effect of separation pressure for a rich condensate stream (SI Units)

Case	Separation Stage Pressures (kPa)	Liquid Produced (m³/h)	Compressor Power Required (kW)
I	8377; 448	55.6	642
II	8377; 3551; 448	56.3	371
III	8377; 3551; 1310; 448	56.5	298

infinity, the lighter molecules are removed preferentially and the recovery of the intermediate components is increased at each stage. The compressor power required is also reduced by stage separation since some of the gas is captured at a higher pressure than would otherwise have occurred [see the example presented in Table 2.1 (oil-field units) and Table 2.2 (SI units)].

2.2.9 Selection of Stages

As more stages are added to the process, there is less and less additional liquid recovery. The incremental income for adding a stage must more than offset the cost of the additional separator, piping, controls, and space and compressor complexities. It is clear that there is an optimum number of stages for each facility. In most cases this number is very difficult to determine, as it may differ from well to well and may change as a well's flowing pressure declines with time. Table 2.3 is a guide to approximating the number of separation stages (excluding the stock-tank), which field experience indicates is somewhat near optimum. This table is meant as a guide and should not replace flash calculations, engineering studies, and engineering judgment.

Through the use of multi-stage separation, the light components are separated from the heavier components. The hydrocarbon liquid leaving each separator is flashed in the next separator at a lower pressure. The amount of liquids recovered in the stock-tank through stage

TABLE 2.3
Stage separation guidelines

Initial Separator Pressure		
kPa	*psig*	*Number of Stages[a]*
170–860	25–125	1
860–2100	125–300	1–2
2100–3400	300–500	2
3400–4800	500–700	2–3[b]

[a]Does not include stock-tank.
[b]At flow rates exceeding 650 m^3/h (100,000 BPD), more stages may be justified.

separation can be calculated by running flash calculations for each stage of separation, including the final flash in the stock-tank. Since the vapor pressure of the liquid leaving the last separator is the same as the separator's operating pressure, there will normally be some vapors that flash in the stock-tank. If enough light components are removed in the multi-stage separators, the heavy components can be stored with a minimum of weathering losses and an increase in liquid recovery.

Crude Oil Treaters

Often gravity separation in three-phase separators is not adequate to separate the water from the crude oil. A common method for separating the emulsion is to *heat the liquid stream*. While this improves the oil–water separation process, it also stabilizes the crude by vaporizing the light hydrocarbons. Quite often this results in higher than desired losses or in crude that has a vapor pressure lower than atmospheric at atmospheric temperatures.

The crude oil volume can be calculated by running a flash calculation at the treater operating condition. For small volumes the oil treating temperature is kept as low as possible to prevent stock-tank losses, since the treated oil will normally go directly to the stock-tank without cooling. For larger volumes the crude leaving the treater is often cooled before going to storage or sale. The crude's vapor pressure can be determined by running a bubble-point calculation on the crude composition leaving the treater at the stock-tank temperature.

Crude Oil Stabilizer

In the preceding section on multi-stage separation, the crude oil was flashed in a separator, light components were removed, and the bubble-point liquid was lowered in pressure. When the pressure is lowered, additional light hydrocarbon vapors are formed and are again removed in the lower-pressure separator. Adding additional stages of separation prior to the crude's going to the stock-tank results in higher liquid recovery and lower loss of hydrocarbon vapor from the stock-tanks.

It is possible to stabilize a hydrocarbon liquid at constant pressure by successively flashing the hydrocarbon liquid at increasing temperatures as shown in Figure 2.3. At each successive stage the partial pressure of the intermediate components is higher than it could have been at that temperature if some of the lighter components had not been removed by the previous stage. It would be very costly to arrange a process as shown in Figure 2.3 and thus never done. Instead, the same effect can be obtained in a tall, vertical pressure vessel with a cold temperature at the top and a hot temperature at the bottom. This unit is called a "stabilizer."

A crude stabilizer utilizes the same principle as multi-stage separation except that the flashes take place in a stabilizer column at a constant pressure, but with varying temperatures. As the hydrocarbon liquid falls from tray to tray in the tower, it is heated by the hot gases bubbling through the liquid. On each tray some of the liquids are vaporized and some of the hot gases are condensed. The liquids falling down the tower become richer in heavy hydrocarbon components and leaner in light hydrocarbons.

By controlling the tower pressure and the bottom temperature, the vapor pressure of the crude oil leaving the bottom of the tower can be controlled. At a set tower pressure, the crude product's vapor pressure can be lowered by increasing the bottom temperature or

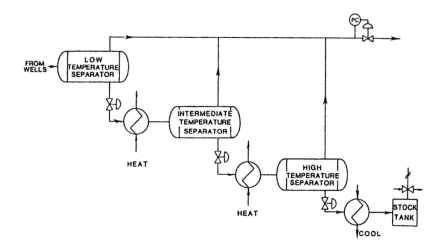

FIGURE 2.3. Multiple flashes at constant pressure and increasing temperature.

can be increased by lowering the bottom temperature. As mentioned previously, the liquid leaving the tower at the bottom tray temperature is in equilibrium with the vapors and is at its bubble-point. The liquid leaving the tower is cooled before going to storage or pipeline. Similarly, the hydrocarbon vapors leaving the top of the stabilizer are in equilibrium with the liquids on the top tray and are at their dew-point.

A stabilizer column separates the lighter components from the heavier components. The stabilizer is normally a trayed tower; however, packing can also be used. As heat is added to the bottom of the tower, vapors are generated on the bottom tray. The hot vapors rise to the tray above, where they bubble through the liquid. The liquid is heated by the hot vapors, which vaporize some of the liquid. The vapors, in turn, are cooled by the liquid, and a portion of the vapor is condensed. This process of vaporization and condensation is repeated on each tray in the tower. As the liquids fall down the tower, the light hydrocarbon components are removed. As the gas goes up the tower, the heavier hydrocarbons are condensed so that the crude oil leaving the tower contains almost none of the light hydrocarbon components, and the vapor leaving the top of the tower contains almost none of the heavier components.

The operating pressure of the stabilizer will normally range from 100 to 200 psi (700–1400 kPa). By operating the stabilizer in this pressure range, the stabilizer's bottom temperature will be in the 200–400 °F (90–200 °C) range. This is within the normal range of heat exchanger operation for fluids used as heating media. For a given bottom product's vapor pressure, a lower stabilizer operating pressure requires a lower bottom temperature, but more compression is required for the overhead vapors.

Another problem that must be considered in the stabilizer design is cold-feed versus reflux. A cold-feed stabilizer without reflux does not achieve as good a split between the light and heavy ends as a column with reflux; therefore, recoveries are not as high. A column with reflux requires additional equipment, higher capital investment, and a higher operating cost, but achieves a higher recovery. Descriptions of a cold-feed stabilizer and a stabilizer with reflux follow.

2.2.10 Cold-Feed Stabilizer

A typical cold-feed stabilization system is shown in Figure 2.4. The crude is being produced to a high-pressure separator operating at 1000 psig (7000 kPa). In this separator, the majority of the very light components, such as methane, can be removed without significant loss of the heavy components, while the bulk produced water is also removed from the crude. The crude oil is then flashed to the stabilizer

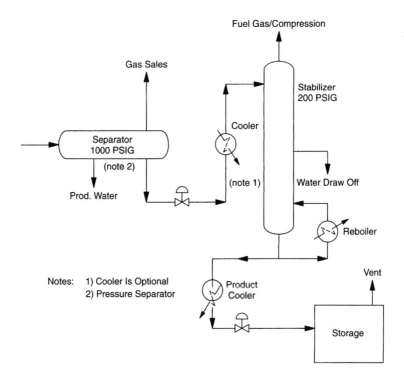

FIGURE 2.4. Schematic of a typical cold-feed stabilization system.

column, which should operate below 200 psig (1400 kPa). The stabilizer is a conventional distillation column with a reboiler but no overhead condenser. The lack of an overhead condenser means that there is no liquid reflux from the overhead stream. Therefore, the feed is introduced on the top tray and must provide all the cold liquid for the tower. Since the feed is introduced on the top tray, it is important to minimize the flashing of the feed so that heavy components are not lost overhead. Sometimes a cooler is included on the feed stream to lower the feed temperature and reduce flashing.

The bottom product from the stabilizer is flashed to storage. The stabilizer should include a product cooler to cool the bottom product and prevent flashing of the liquid as it is lowered to storage pressure. Depending on operating pressure, the bottom temperature will range from 90 to 200 °C (200–400 °F).

The overhead gas can be used as fuel or compressed and included with the sales gas. Any water that enters the column in the feed stream will collect in the middle of the column due to the range of temperatures involved. This water cannot leave with the bottom product or with the overhead stream; therefore, provisions should be made to remove this water from a tray near the middle of the column. The heating of the crude in the tower acts as a demulsifier to remove water from the crude. The excellent water-separating ability of the stabilizer usually eliminates the need for a crude dehydration system.

2.2.11 Crude Stabilizer with Reflux

A typical crude stabilization system with reflux and a feed/bottom heat exchanger is shown schematically in Figure 2.5. In this application the column's top temperature is controlled through cooling and condensing part of the hydrocarbon vapors leaving the tower and pumping the resulting liquids back to the tower. This replaces the cold-feed to the previous stabilizer and allows better control of the overhead product and, consequently, slightly higher recovery of the heavier ends.

Since the reflux condenser provides the cooling for the stabilizer top, cold-feed is no longer required. As shown, the feed to the stabilizer is heated by exchanging heat with the stabilizer's bottom product. The feed is now introduced into the middle section of the stabilizer.

The principles of the stabilizer are the same as those discussed previously. Detailed selection of a stabilizer system is outside the scope of this tutorial and requires economic justification. This selection is best done by an engineer familiar with the design of crude stabilization systems. The added equipment costs of one scheme must be evaluated against its potentially higher crude oil recovery.

A heat balance around the tower is part of the design. The heat leaves the tower in the form of vapors out the top, and the liquid bottom product has to be balanced by the heat entering in the feed and

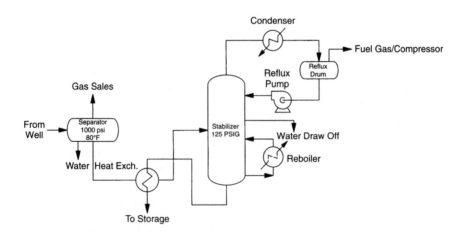

FIGURE 2.5. Schematic of a typical crude stabilization with reflux and feed/bottom heat exchanger.

the reboiler. Naturally, if the column has a reflux, this amount of heat has to be added to the column balance.

A properly designed and operated stabilizer can meet a desired crude oil vapor pressure and results in product recovery superior to that of a typical multi-stage separator system, but the initial investment and operation costs of a stabilizer will be higher.

2.2.12 Key Components

Vapor pressure alone does not define the ease of separation of the components from liquid mixtures, because the vapor pressure of each component is affected by the composition of the liquid mixture. The pure component vapor pressure of a component divided by the pure component vapor pressure of another component at the same temperature is a direct measure of the ease of separating these two components. This ratio of pure component vapor pressures is called the "relative volatility" of the two components. Relative volatilities of similar components generally vary only slightly with temperature.

Since relative volatilities indicate the ease of separating two components, they are also a measure of the percentage of each component that will separate into the vapor state. Therefore, selecting a particular component and designing a system that creates a specific separation between vapor and liquid for this component will fix the degree of separation for each component in the mixture. This will determine the compositions of the top and bottom streams.

Key components in a multi-component distillation are those whose relative volatilities are such that they determine the component split into the vapor and liquid phases. These components tend to appear in both the vapor and liquid phases while other components will appear largely in either one phase or the other. The key component appearing mostly in the vapor phase is called the "light key," while the key component in the liquid phase is termed the "heavy key." All other components are termed "non-keys." For crude stabilizers, it is convenient to use normal butane (nC_4) or iso-pentane (iC_5) as the heavy key and propane (C_3) or iso-butane (iC_4) as the light key.

2.2.13 Reid Vapor Pressure

Reid vapor pressures (RVPs) are sometimes specified by crude oil purchasers, particularly if the crude is to be transported by tanker or truck prior to reaching a processing plant. Purchasers specify low RVPs so that they will not be paying for light components in the liquid, which will be lost due to weathering.

RVP is used to characterize the volatility of gasolines and crude oils. The RVP of a mixture is determined experimentally according to a procedure standardized by the American Society for Testing Materials at 100 °F (37.8 °C). A sample is placed in a container such that the ratio of the vapor volume to the liquid volume is 4 to 1. The absolute pressure at 100 °F (37.8 °C) in the container is the RVP for the mixture.

Since a portion of the liquid has been vaporized to fill the vapor space, the liquid has lost some of the lighter components. This effectively changes the composition of the liquid and thus yields a slightly lower vapor pressure than the true vapor pressure of the liquid in its original composition. The RVP of a mixture, then, is slightly lower than the true vapor pressure of the mixture at 100 °F (37.8 °C). The GPSA gives the approximate true vapor pressure for gasolines and crude oils from their RVP at various temperatures.

Despite the inaccuracy of RVP measurement, it is used to specify volatility limits for crude oils in sales contracts. Crude stabilization systems can be designed to meet RVP requirements because the RVP of a mixture is always less than the true vapor pressure at 100 °F (37.8 °C). Therefore, the separation in the stabilizer should be designed to yield a mixture with a true vapor pressure at 100 °F (37.8 °C) equal to the RVP requirement. This will yield a bottom product with an RVP slightly below the required RVP. Typical RVP requirements for crude sales range from 10 to 12 psi (70 to 82 kPa) RVP.

2.2.14 Component Recovery

When crude RVP is not specified, then some particular component split may be specified. The particular specification may be chosen for a variety of reasons. Some of the possible specifications are:

1. A percentage recovery of the heavy key in the liquid.
2. A maximum mole percentage of the light key in the liquid.
3. A maximum mole percentage of the heavy key in the gas.

Specification 1 can control the process to achieve the most economical split between gas and liquid. Specification 2 limits the amount of light components lost from the gas and limits the liquid vapor pressure. Specification 3 limits the amount of heavy components lost to the gas and limits the gas heating value.

2.2.15 Column Constraints

In most cases when a crude stabilizer is used, the process and product compositions are controlled by setting constraints on the column operating conditions. For example, to obtain a crude with a specified RVP, the pressure and the bottom temperature are determined. Similarly, to obtain a specified maximum amount of the heavy key in the gas, the pressure and overhead temperature must be determined.

2.3 Equipment Description

2.3.1 Stabilizer Column

The stabilizer column is a fractionation tower using trays or packing, as shown in Figure 2.6. Trays, structured packing, or random packing in the column are used to effect an intimate contact between the vapor and liquid phases, permitting the transfer of mass and heat from one phase to the other. The trays are orifice-type devices designed to disperse the gas uniformly on the tray and through the liquid on the tray. Trays are commonly spaced 24 in. apart. The three most widely used trays are the valve, bubble cap, and perforated. Standard random packing, available in numerous sizes, geometric shapes, and materials, is composed of solids randomly packed in the tower. Structured packing is made of folded perforated plates welded together. The trays shown in Figure 2.6 are bubble cap. Figures 2.7, 2.8 and 2.9 show trays and packing in more detail.

In general, bubble cap trays are preferable for a stabilizer column. Structured packing can be used to decrease both the height and the

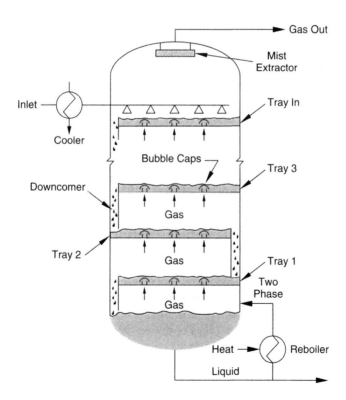

FIGURE 2.6. Schematic of stabilizer tower.

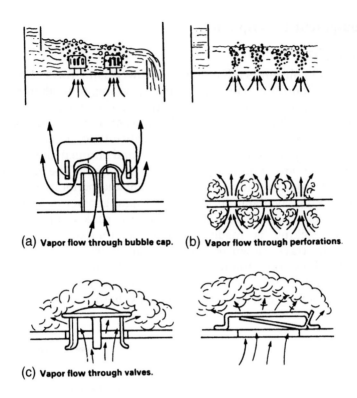

(a) **Vapor flow through bubble cap.**   (b) **Vapor flow through perforations.**

(c) **Vapor flow through valves.**

FIGURE 2.7. Vapor flow through trays.

diameter of the stabilizer column, but it is more costly than bubble caps and more likely to become plugged. Random packing is commonly used in small-diameter columns [less than 50 cm (20 in.)] because of the difficulty of installing bubble caps in these columns. Stabilizers may use from 5 to 50 trays, but 10 to 12 trays are most common and are usually sufficient.

The feed to the tower will normally enter the tower near the top of a cold-feed stabilizer and at or near the tray where the tower conditions and feed composition most nearly match the inlet feed conditions, in towers with reflux. The liquids in the tower fall down through the down-comer, across the tray, over the weir, and into the down-comer to the next tray. The temperature on each tray increases as the liquids drop from tray to tray. Hot gases come up the tower and bubble through the liquid on the tray above, where some of the heavier ends in the gas are condensed and some of the lighter ends in the liquid are vaporized. The gas gets leaner and leaner in heavy hydrocarbons as it goes up the tower; the falling liquids become richer and richer in the heavier hydrocarbon components.

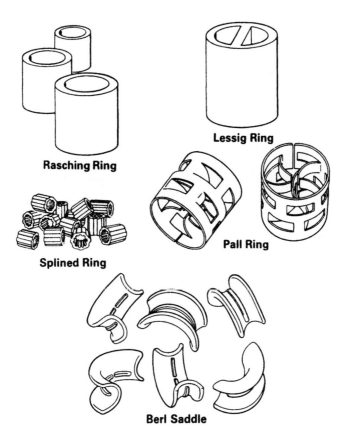

Lessig Ring

Rasching Ring

Splined Ring

Pall Ring

Berl Saddle

FIGURE 2.8. Various types of random packing.

The vapors leaving the top of the tower contain a minimum amount of heavy hydrocarbons, and the liquid leaving the bottom of the tower contains a minimum of light hydrocarbons. The stabilizer column normally operates at a pressure of 100–200 psig (700–1400 kPa).

The process design calculations for a crude stabilizer are very complex and are not usually performed by hand. Computer process simulation programs for multi-component distillation can be used to model the operation of crude stabilizers. These programs can aid in determining the number of trays required, internal flow rates from tray to tray, operating temperatures, operating pressures, product compositions, and heat balances. The speed of these programs allows the designer to quickly investigate changes in feed temperature, cold-feed stabilization versus reflux, number of trays, etc., and thus arrive at the most economic design.

FIGURE 2.9. Structured packing can offer better mass transfer than trays (Courtsey of Koch Engineering Co. Inc.)

2.3.2 Stabilizer Reboiler

The stabilizer reboiler boils the bottom product from the column as in other distillation processes. The reboiler is the source for all the heat used to generate vapor in a crude stabilizer. When we control the heat input with the reboiler, the boiling point of the bottom product can be controlled. Together with the stabilizer operating pressure, this action controls the vapor pressure of the bottom product.

The reboiler may be either a kettle-type or a thermosiphon-type reboiler. Typically, reboiler temperatures range from 200–400 °F (90–200 °C) depending on operating pressure, bottom product composition, and vapor pressure requirements. Reboiler temperatures should be kept to a minimum to decrease the heat requirements, limit salt build-up, and prevent corrosion problems.

When stabilizer operating pressures are kept below 100 psig (700 kPa), the reboiler temperatures will normally be below 300 °F (150 °C). A water–glycol heating medium can then be used to provide heat. Higher stabilizer operating pressures require the use of steam or hydrocarbon-based heating mediums. Operating at higher pressures, however, decreases the flashing of the feed on entering the column, decreasing the amount of feed cooling required. In general, a crude stabilizer should be designed to operate between 100 and 200 psig (700 and 1400 kPa).

Selection of a stabilizer heat source depends on the heating medium and column operating pressure. The source of reboiler heat should be considered when a crude stabilizer is being evaluated. If turbine generators or compressors are installed nearby, then waste heat recovery should be considered. In addition, or alternately, fired heaters should be investigated. These factors must be considered when designing a crude stabilization system.

2.3.3 Stabilizer Bottom Product Cooler

The stabilizer bottom product cooler is used to cool the bottom product leaving the tower before it goes to a tank or pipeline. Normally, the bottom product will be cooled from the operating temperature of 150–200 °C (300–400 °F) to 40–50 °C (100–120 °F). The temperature of the bottom product may be dictated by sales contract or by efforts to prevent loss of vapors from an atmospheric storage tank.

The cooler may be an air-cooled exchanger or a shell-and-tube exchanger, which is quite often used when there is another system that needs to be heated. For a stabilizer with a reflux system, the bottom product may be cooled by exchanging heat with the feed to the stabilizer. The presentation on heat exchangers, covered in Volume 2 of the "Surface Production Operations"[1] series, provides procedures for the selection and sizing of shell-and-tube heat exchangers.

2.3.4 Stabilizer Reflux System

A stabilizer reflux system consists of a reflux condenser, reflux accumulator, and reflux pumps. The system is designed to operate at a temperature necessary to condense a portion of the vapors leaving the top of the stabilizer. The temperature range can be determined by calculating the overhead vapor's dew-point temperature. The heat duty required is determined by the amount of reflux required.

Selection of the type of exchanger for the reflux condenser depends upon the design temperature required to condense the reflux. The lower the operating pressure of the stabilizer, the lower the temperature required for condensing the reflux will be. In most installations, air-cooled exchangers may be used. Other instances may require refrigeration, and a shell-and-tube-type exchanger will be used.

The reflux accumulator is a two-phase separator with several minutes of retention time to allow separation of the vapors and liquids. The reflux accumulator is normally located below the reflux condenser, with the line sloped from the condenser to the accumulator. The reflux accumulator must be located above the reflux pumps to provide the necessary net positive suction head (NPSH) required by the pumps. The size of the reflux accumulator depends on the amount of reflux required and the total amount of vapors leaving the tower.

Reflux pumps are sized to pump the required reflux from the reflux accumulator back to the top of the stabilizer. Normally, these pumps are designed with a delta pressure of 50 psi (340 kPa). Depending upon the reflux circulation rate, two 100% pumps or three 50% pumps may be installed. This allows either a 100% spare or a 50% spare pump.

2.3.5 Stabilizer Feed Cooler

For a cold-feed stabilizer, an inlet feed cooler may be required. Again, calculations are required to determine the design feed temperature and the heat duty of the exchanger. This exchanger is usually a shell-and-tube type with some type of refrigerant required to cool the feed sufficiently.

The selection of equipment and the decision on a cold-feed versus a reflux stabilizer depend on several considerations. The availability of heat sources for the reboiler and streams for cooling the system influences the final decision. As usual, the economics of product recovery, capital investment, and operation costs will be major factors.

2.3.6 Stabilizer Feed Heater

For stabilizers with a reflux system, a feed heater may be required. If a feed heater is used, it is normally a shell-and-tube type that exchanges heat between the cold-feed and the hot bottom product, which is then cooled before going to storage.

2.4 Design Procedure

The objectives of the design procedure are to estimate and analyze the liquid recovery by comparing stage separation with a stabilizer. The procedure in the upcoming section on multi-stage separator design is adequate to estimate the recovery and the primary operating ranges and equipment sizes for a crude stabilizer.

For detailed design and evaluation, once the wellstream fluid compositions are known, the calculations should be simulated on a computer. Most computer simulations generate material and energy balances for the process, compare different inlet temperatures, and evaluate recoveries with and without column reflux. From this information, the size of the tower, reboiler duty, cooler duty, and condenser duty can be determined and preliminary cost estimates made.

For the preliminary calculations the following information is required:

- production rate,
- Wellstream composition,
- flowing wellstream temperature, and
- *Product specifications:* vapour pressure or percentage of key component allowed in product streams.

2.4.1 Multi-Stage Separator Design

Based on wellstream composition and producing condition, assume separator operating pressures and the number of separator stages. Factors such as gas sales pressure and existing compressor suction pressures, if any, need to be evaluated when estimating the separator operating pressure. If flowing wellhead pressures are high enough, the high-pressure separator will normally operate at a pressure slightly higher than gas sales line pressure. Through the use of the computer, a number of different options can be evaluated and recoveries calculated.

2.4.2 Multi-Stage Separator Design Procedure

The following steps outline the procedure to use when designing a multi-stage separator:

1. At the high-pressure separator's operating pressure and temperature, read K values from the GPSA section on "Equilibrium Ratio (K) Data" for each component in the wellstream. For a normal separation pressure of 1000 psia (6900 kPa), use a convergence pressure of 3000 psi (20,700 kPa) or calculate

the convergence pressure in accordance with the method given in the GPSA data book. Note that K value charts are not available for CO_2. The GPSA data book recommends that the K value for CO_2 be approximated as the square root of the K value for methane times the K value for ethane.

2. Since this is a trial-and-error solution, estimate the moles of liquid leaving the separator and calculate V/L, where V is moles of vapor and L is moles of liquid.

3. Calculate the moles of liquid for each component by using the equation:

$$L_N = \frac{F_N}{K_N\left(\dfrac{V}{L}\right) + 1}.$$

4. Obtain the total moles of gas and liquid by adding the moles of liquid and gas for each component. Compare the calculated V/L and repeat steps 2 and 3 until the assumed V/L and the calculated V/L are equal. If the calculated V/L is higher than the assumed V/L, for the next assumption the V/L should be increased. Depending on the composition of the liquid and the pressure drop, the temperature may decrease when the liquid is dropped in pressure from the high-pressure separation to the next stage.

5. At this separation stage, read the K values for each component from the GPSA data book for the assumed operating pressure and temperature, using a convergence pressure of 2000 psia (13,800 kPa). Using the moles of liquid from the high-pressure separator, again calculate the moles of vapor and of liquid leaving the separator following the previous procedure.

6. Repeat the flash calculations for each stage of separation, including the stock-tank, removing the gas from each separation stage and using the liquids from the previous separator. The total number of moles remaining as liquid after the last flash represents the volume and composition of the liquid product obtained through stage separation.

7. The crude oil vapor pressure will be the same as the pressure of the final flash calculation. Assuming this is a stock-tank at 14.7 psia (101.4 kPa) and 100 °F (37.8 °C), then the vapor pressure is 14.7 psia (101.4 kPa). If a sales product vapor pressure less than 14.7 psia (101.4 kPa) is necessary and multistage separators are used, the crude may require heating above 100 °F (37.8 °C) prior to entering the low-pressure separator.

8. To calculate the required amount of heat, estimate a temperature at the low-pressure separator. The higher the operating pressure of the separator, the higher the temperature necessary to lower the stock-tank's vapor pressure will be. Using the assumed temperature and pressure, run a flash calculation to determine the composition of the liquid. This liquid must be cooled before leaving the separator to go to the stock-tank.
9. Using the liquid composition, calculate the liquid's bubble-point pressure at stock-tank conditions through use of the equation: $\Sigma K_N F_N = \Sigma F_N$. Assume a vapor pressure at stock-tank temperature, read the individual component K values, and calculate the liquid's bubble-point. Repeat this procedure until the assumed pressure equals the calculated bubble-point pressure.
10. If the calculated bubble-point pressure is higher than the specified sales vapor pressure, a higher separation temperature is assumed and the calculations repeated until the desired vapor pressure is reached. If the bubble-point pressure is lower than the specified sales product's vapor pressure, the temperature assumed is too high and a lower temperature must be assumed and calculations repeated as before.
11. Separator sizes can be calculated for each separation stage using the principles in the tutorials on two-phase separators and three-phase separators. The cost of the separators can be estimated based on these sizes.

2.4.3 Crude Oil Treater Design

Crude oil treaters are included as a form of stabilization since it may be necessary to heat the crude to remove the water. Heating the crude prior to separation will often result in a loss of product and, consequently, a lower vapor pressure in the final product.

If it appears that a crude oil treater will be required, the crude oil treating temperature should be estimated. Using this temperature and the composition of the liquid entering the treater, the liquid outlet composition can be calculated through flash calculations.

By using this liquid composition and the stock-tank's temperature, the liquid's bubble-point pressure can be calculated. This is the same as the vapor pressure of the crude oil at stock-tank conditions. If the calculated bubble-point is higher than the specified sales vapor pressure, a higher treating temperature or some other form of stabilization may be required. Most crude oil streams are treated at a temperature high enough to eliminate the problem of product vapor pressure being too high.

2.4.4 Crude Oil Stabilizer Design

The calculations involving crude oil stabilizer design are much more complicated and should be done using a suitable computer program.

References

[1] Arnold K. E., and Stewart M. I., *Surface Production Operations—Design of Gas Handling Facilities*, 2nd Edition, Gulf publishing, 1993.

CHAPTER 3

Produced Water Treating Systems

3.1 Introduction

When hydrocarbons (crude oil, condensate, and natural gas) are produced, the wellstream typically contains water produced in association with these hydrocarbons. The produced water is usually brine, brackish, or salty in quality but in rare situations may be nearly "fresh" in quality. The water must be separated from the hydrocarbons and disposed of in a manner that does not violate established environmental regulations. Typically, the produced water is mechanically separated from the hydrocarbons by passing the wellstream through process equipment such as three-phase separators, heater-treaters, and/or a free-water knockout vessel. These mechanical separation devices do not achieve a full 100% separation of the hydrocarbons from the produced water. The produced water separated from the hydrocarbons in these mechanical separation devices will contain 0.1–10 vol.% of dispersed and dissolved hydrocarbons. Produced water treating facilities are used to further reduce the hydrocarbon content in the produced water prior to final disposal.

Regulatory standards for overboard disposal of produced water into offshore surface waters vary from country to country. Failure to comply with such regulations can often result in civil penalties, large fines, and lost or deferred production. Intentional violation of these regulations can result in criminal prosecution of officers and other individuals acting on behalf of the company who intentionally neglected compliance. Currently, regulations require the "total oil and grease" content of the effluent water to be reduced to levels ranging between 15 and 50 mg/l depending upon the host country. For U.S. offshore operations, the current standard is 29 mg/l.

Disposal of produced water into onshore surface waters is generally prohibited by environmental regulations. Onshore disposal typically requires the produced water effluent to be injected into a saltwater disposal well. Onshore produced water treating for hydrocarbon removal prior to injection into a saltwater disposal well is not commonly regulated; however, a regulatory permit is typically required before initiating any substrata water disposal injection project. The permitting procedure helps to safeguard subsurface drinking water supplies by assuring that disposal wells are drilled and completed in a manner so the fresh drinking water supply zones are isolated from communication with any brackish or salty water zones.

The purpose of this chapter is to present the engineer with a procedure for selecting the appropriate type of equipment for treating oil from produced water and to provide the theoretical equations and empirical rules necessary to size the equipment. When this design procedure is followed, the engineer will be able to develop a process flow sheet, determine equipment sizes, and evaluate vendor proposals for any wastewater treating system once the discharge quality, the produced water flow rate, the oil specific gravity, the water specific gravity, and drainage requirements are determined.

3.2 Disposal Standards

3.2.1 Offshore Operations

Standards for the disposal or produced water to surface waters both onshore and offshore are developed by governmental regulatory authorities. Table 3.1 summarizes offshore disposal standards for several countries. The standards are current as of this writing.

In addition to placing limits on the oil content, regulatory agencies generally specify an analytical method for determining the oil content. A number of analytical methods are available, and they

TABLE 3.1
Worldwide produced water effluent oil concentration limitations

Ecuador, Colombia, Brazil	30 mg/l All facilities
Argentina and Venezuela	15 mg/l New facilities
Indonesia	25 mg/l Grandfathered facilities
Malaysia, Middle East	30 mg/l All facilities
Nigeria, Angola, Cameroon, Ivory Coast	50 mg/l All facilities
North Sea, Australia	30 mg/l All facilities
Thailand	50 mg/l All facilities
USA	29 mg/l OCS water
	Zero discharge inland water

produce different amounts of oil measured and reported for the same sample. Analytical methods are discussed in Appendices A–C.

Produced water toxicity is regulated only in the United States, where a government permit is necessary to limit the toxicity of produced water discharged into the waters.

3.2.2 Onshore Operations

Disposal of produced water into freshwater streams and rivers is generally prohibited except for the very limited cases where the effluent is low in salinity. Some oil-field brines might kill freshwater fish and vegetation due to high salt content.

Regulatory agencies generally require that produced water from onshore operations be disposed of by subsurface injection, although there are limited exceptions. In addition to requiring subsurface disposal, regulatory agencies regulate the completion and operation of the disposal wells.

3.3 Characteristics of Produced Water

Produced water contains a number of substances, in addition to hydrocarbons, that affect the manner in which the water is handled. The composition and concentration of substances may vary between fields and even between different production zones within a single field. The terminology used for concentration is milligrams per liter (mg/l), which is mass per volume ratio and is approximately equal to parts per million (ppm). Some of the important produced water constituents are discussed in this section.

3.3.1 Dissolved Solids

Produced waters contain dissolved solids, but the amount varies from less than 100 to over 300,000 mg/l, depending on the geographical location as well as the age and type of reservoir. In general, water produced with gas is condensed water vapor with few dissolved solids and will be fresh with a very low salinity. Aquifer water produced with gas or oil will be much higher in dissolved solids. A relationship exists between the reservoir temperature (which is related to subsurface depth) and the amount of total dissolved solids (TDS). Produced water from hot reservoirs tends to have higher TDS concentrations while cooler reservoirs tend to have lower levels of TDS.

Dissolved solids are inorganic constituents that are predominantly sodium (Na^+) cations and chloride (Cl^-) anions. Other common cations are calcium (Ca^{2+}), magnesium (Mg^{2+}), and iron (Fe^{2+}), while barium (Ba^{2+}), potassium (K^+), strontium (Sr^+), aluminum (Al^{3+}), and lithium (Li^+) are encountered less frequently. Other anions present are bicarbonate (HCO_3^-), carbonate (CO_3^{2-}), and sulfate (SO_4^-).

All produced water treating facilities should have water analysis data for each major reservoir and for the combined produced water stream. Especially important are constituents that could precipitate to form scales.

3.3.2 Precipitated Solids (Scales)

The more troublesome ions are those that react to form precipitates when pressure, temperature, or composition changes occur. These are the well-known deposits that form in tubing, flowlines, vessels, and produced water treating equipment.

Mixing of oxygenated deck drain water with produced water should be avoided because this may result in the formation of calcium carbonate ($CaCO_3$), calcium sulfate ($CaSO_4$), and iron sulfide (FeS_2) scale, along with oil-coated solids.

Calcium Carbonate ($CaCO_3$)

Calcium carbonate ($CaCO_3$) precipitate can be formed by mixing two dissimilar waters, but the usual cause is the reduction in pressure and release of dissolved carbon dioxide from produced water. This increases the produced water's pH, which reduces the solubility of $CaCO_3$ and leads to scale precipitate. Temperature effects are equally important since $CaCO_3$ is less soluble at higher temperatures and will form a deposit in heat exchangers, heaters, and treaters. Its solubility is approximately 1000 mg/l at 60 °F (15 °C) and diminishes to 230 mg/l as temperature is increased to 200 °F (93 °C). Fortunately, higher salinity increases $CaCO_3$ solubility in produced water to a value greater than that given above for pure water.

Calcium Sulfate ($CaSO_4$)

Calcium sulfate ($CaSO_4$) is one of several sulfate scales and is also called gypsum. Like $CaCO_3$, it can form either as a result of mixing dissimilar waters or naturally as a result of changes in temperature and pressure as the water travels from the subsurface to the surface treating facility. $CaSO_4$ solubility is at its maximum level of 2150 mg/l at approximately 100 °F (38 °C) and diminishes to 2000 mg/l as it cools to 60 °F (15 °C). The solubility of $CaSO_4$ also declines with increasing temperature above 100 °F with its solubility reducing to 1600 mg/l at 200 °F (93 °C). $CaSO_4$ also increases in solubility as the salinity of the produced water increases.

Iron Sulfide (FeS$_2$)

Iron sulfide (FeS$_2$) is a product of corrosion caused by waters containing dissolved hydrogen sulfide coming into contact with equipment fabricated from carbon steel or iron materials. Mixing water containing iron cations (Fe^{2+}) with another water containing hydrogen sulfide will also result in an FeS$_2$ precipitate.

Barium and Strontium Sulfate (BaSO$_4$ and SrSO$_4$)

Barium and strontium sulfate (BaSO$_4$ and SrSO$_4$) are much less soluble than calcium sulfate, but they are not as common in produced waters. BaSO$_4$ solubility is quite low, having a value of approximately 3 mg/l over the range from 100 (15 °C) to 200 °F (93 °C). SrSO$_4$ solubility is 129 mg/l at 77 °F (25 °C) and diminishes to 68 mg/l as the solution temperature increases to 257 °F (125 °C). If a produced water stream containing appreciable quantities of barium or strontium ions is mixed with sulfate-rich water, barium and/or strontium scaling can be expected. These waters are incompatible due to this scaling characteristic and should not be mixed.

3.3.3 Scale Removal

Hydrochloric acid can be used to dissolve calcium carbonate and iron sulfide scales. However, iron sulfide chemically reacts with hydrochloric acid and produces hydrogen sulfide, a highly toxic gas having the odor of rotten eggs. Due to the high toxicity of hydrogen sulfide, safety provisions need to be implemented.

Calcium sulfate is not soluble in hydrochloric acid, but chemicals are available that will convert it to an acid-soluble form that can then be removed by the acid. This process is slow, however, because a two-step process must be repeated to strip the scale layer by layer. Thus, the removal of calcium sulfate is more difficult than the removal of calcium carbonate.

Practical means of dissolving barium or strontium sulfate are not available. These hard scales can be removed by mechanical means, which is a time-consuming process. Mechanical removal of scale can create a disposal problem for the resulting waste material and possibly could result in contamination by naturally occurring radioactive materials (NORM).

3.3.4 Controlling Scale Using Chemical Inhibitors

Scale-inhibiting chemicals are available to retard or prevent all types of scale. They mostly function by enveloping a newly precipitated crystal, thereby retarding growth. Common scale inhibitors include:

- inorganic phosphates (inexpensive but only applicable at low temperature),
- organic phosphate esters (easy to monitor but limited to temperatures below 100 °F),
- phosphates (easy to monitor and have a higher thermal salinity to 150 °F), and
- polymers (best thermal stability and effectiveness, but difficult to monitor).

3.3.5 Sand and Other Suspended Solids

In addition to scale particles, produced water often contains other suspended solids. These include formation sand and clays, stimulation (fracturing) proppant, or miscellaneous corrosion products. The amount of suspended solids is generally small unless the well is producing from an unconsolidated formation, in which case large volumes of sand can be produced. Produced sand is often oil wet and its disposal is a problem. Sand removal is discussed in a later subsection entitled "Equipment Description" under Section 3.6.

Small amounts of solids in produced water may or may not create problems in water treating depending on the particle micron size and its relative attraction to the dispersed oil. If the physical characteristics and electronic charge of such solids result in an attraction to the dispersed oil droplets, the solid particles can attach to the dispersed oil droplets to stabilize emulsions, thereby preventing coalescence and separation of the oil phase. The combined specific gravity of the resulting oil/solid droplet can be approximately equal to that of the produced water, and gravity separation becomes difficult if not impossible.

The concentration of suspended solids can be monitored with a 0.45-µm Millipore filter test, and residue can be analyzed for mineral content in an attempt to identify the source of the solid.

When solids are present, the following practices should be applied:

- Chemical treatment must be used to "break" the electronic attraction between the solid particle and the oil droplet.
- Equipment design must incorporate solids removal ports, jets, and/or plates.
- Oil measurement techniques not affected by the solids should be used.
- Solids are likely to be oil coated, and offshore disposal may be prohibited, as is the case in the United States. This applies to solids removed from desanders or vessels, not to the solids suspended in the water.

- Water injection for disposal should be made into a disposal zone that has a high enough permeability to prevent the suspended solids from plugging. Consideration should be given to using filtration equipment to remove the larger particles prior to injection into the disposal well. Periodic back flowing and acidizing are generally needed to maintain disposal wells if filtration is not applied.
- Water-flood injection for pressure maintenance and additional recovery often requires filtration (to remove suspended solids). Water injection pressures typically must be maintained at pressure levels below the fracture gradient pressure of the formation.

3.3.6 Dissolved Gases

The most important gases found in produced water include natural gas (methane, ethane, propane, and butane), hydrogen sulfide, and carbon dioxide. In the reservoir the water can be saturated with these gases at relatively high pressures. As the produced water flows up the wells, most of these gases flash to the vapor phase and are removed in primary separators and stock tanks. The pressures and temperatures at which the produced water is separated from the main oil, condensate, and/or natural gas streams will impact the quality of dissolved gas that will be contained in the produced water stream feeding the water treating facilities. The higher the separation pressure, the higher the quantity of dissolved gases will be. An inverse relationship holds true for the effects of temperature: the higher the separation temperature, the lower the quantity of dissolved gases will be.

Natural gas components are slightly soluble in water at moderate to high pressures and will be present in the produced water stream. The solubility of natural gas (primarily methane) is a function of pressure, temperature, and specific gravity of the water. It is interesting to note that natural gas components have an affinity for the dispersed oil droplets, and this principle is applied to the design of gas flotation equipment commonly used in produced water treating systems.

If hydrogen sulfide is present in the produced reservoir fluid, or if sulfate reducing bacteria are a problem in the reservoir or production equipment, hydrogen sulfide will likewise be present in the produced water stream. Hydrogen sulfide is corrosive, can cause iron sulfide scaling, and is extremely toxic if inhaled. The toxicity of hydrogen sulfide hinders operation and maintenance of equipment, particularly when the vessels must be opened for adjustments, as in the case when weir adjustments are required in gas flotation cells. Special training and life support breathing equipment are recommended for use by personnel when such activities result in exposure to hydrogen sulfide. Additionally, iron sulfide (the corrosion product of hydrogen sulfide) presents a potential fire hazard since it is prone to auto-ignition when exposed to air or other sources of oxygen.

If carbon dioxide is present in the produced reservoir fluid, it too will be present in the produced water. Carbon dioxide is corrosive and can cause $CaCO_3$ scaling. On the other hand, removal of CO_2 and H_2S will result in increased pH, which could lead to scaling.

Oxygen is not found naturally in produced water. However, when the produced water is brought to the surface and exposed to the atmosphere, oxygen will be absorbed into the water. Water containing dissolved oxygen can cause severe and rapid corrosion, solids generation from oxidation reactions, and oil weathering that inhibits cleanup. To prevent this, a natural gas blanket should be maintained on all of the production and water treating tanks and vessels used within the process.

Seawater is often used as the source of water for water floods and water injection pressure maintenance projects offshore. Seawater contains considerable amounts of dissolved oxygen and some carbon dioxide. Bacteria in untreated seawater may also be a problem. The oxygen and carbon dioxide must be removed from the source water by either vacuum de-aeration or gas stripping prior to injection.

3.3.7 Oil in Water Emulsions

Most emulsions encountered in the oil field are water droplets in an oil continuous phase and are called "normal emulsions." The water is dispersed in the form of very small droplets ranging between 100 and 400 μm in diameter. Oil droplets in a water continuous phase are known as "reverse emulsions" and can occur in produced water treating operations.

If the emulsion is unstable, the oil droplets will coalesce when they come in contact with each other and form larger droplets, thus breaking the emulsion. An unstable emulsion of this type will break within minutes.

A stable emulsion is a suspension of two immiscible liquids in the presence of a stabilizer or emulsifying agent that acts to maintain an interfacial film between the phases. Chemicals, heat, settling time, and electrostatics are used to alter and remove the film and cause emulsion breakdown. Untreated stable emulsions can remain for days or even weeks.

Emulsion breakers for water-in-oil emulsions, also known as destabilizers or demulsifiers, are oil-soluble and are added to the total well stream ahead of the process equipment. Being oil-soluble, the emulsion breaker is carried with the crude. Thus, if the emulsion is not broken in the first-stage separator, the chemical has additional time to act in the subsequent separators and the stock tank.

Oil-in-water emulsions can be broken by "reverse emulsion breakers," which are special destabilizers or demulsifiers. These are similar to the conventional emulsion breakers except that they are water-soluble. Reverse emulsion breakers are generally injected into the water stream after the first oil–water separation vessel. Typical concentrations are in the 5- to 15-ppm range, and over-treating should be avoided because these chemicals can stabilize an emulsion.

The emulsions in produced water will become oil in the form of dispersed droplets after the emulsion film is broken. The droplets will coalesce to yield an oily film that can be separated from the produced water using gravity settling devices such as skim vessels, coalescers, and plate separators. However, small droplets require excessive gravity settling time, so flotation cells or acceleration enhanced methods such as hydrocyclones and centrifuges are used. Equipment selection is based on the inlet oil's droplet diameter and concentration.

3.3.8 Dissolved Oil Concentrations

Dissolved oil is also called "soluble oil," representing all hydrocarbons and other organic compounds that have some solubility in produced water. The source of the produced water affects the quantity of the dissolved oil present. Produced water derived from gas/condensate production typically exhibits higher levels of dissolved oil. In addition, process water condensed from glycol regeneration vapor recovery systems contains aromatics including benzene, toluene, ethyl benzene, and xylenes (BTEX) that are partially soluble in produced water.

Gravitational-type separation equipment will not remove dissolved oil. Thus, a high level of total oil and grease could be discharged if the produced water source contains significant quantities of dissolved oil. Produced water streams containing high concentrations of dissolved oil can be recycled to a fuel separator to help reduce the quantity of dissolved oil in the water effluent. Other technologies, such as bio-treatment, adsorption filtration, solvent extraction, and membranes, are currently being evaluated by the industry for removing dissolved oil, but such processes are not yet readily available for commercial applications.

It is essential that actual water test analysis data for dissolved and dispersed oil concentrations are needed in the planning stage prior to designing a water treating facility for a specific application. If the design engineer assumes a value for the dissolved oil content without first having obtained actual water test analysis for the specific produced water stream to be treated, the facility design may not be capable of treating the effluent water to meet regulatory compliance specifications. Therefore, lab testing is required first.

The solubility of crude oil in produced water has not been extensively documented, but the solubility of several hydrocarbons can illustrate the potential range. Field experience indicates that solubility does not change appreciably with the temperatures used during water treating, specifically from 77–167 °F (25–75 °C). Solubility does increase significantly, however, as temperatures rise above 167 °F (75 °C).

The effect of high salinity on reducing the solubility of dissolved hydrocarbons implies that produced water from gas well and gas processing sources should be mixed with the saltiest brine available to reduce the dissolved oil concentration. The dissolved hydrocarbons would be forced out of solution from the water into the vapor phase or into a dispersed oil droplet removed by gravity separation equipment.

Water chemistry and hydrocarbon solubility are also related to toxicity. Dissolved saturated paraffinic (aliphatic) petroleum hydrocarbons have low solubilities in water and have not demonstrated toxicity. Aromatics, such as benzene, toluene, ethyl benzene, and xylene, are more soluble and more toxic.

3.3.9 Dispersed Oil

Dispersed oil can consist of oil droplets ranging in size from about 0.5 μm in diameter to greater than 200 μm in diameter. The oil droplet size distribution is one of the key parameters influencing the produced water treating performance. According to Stokes' law, the rising velocity of an oil droplet is proportional to the square of the droplet diameter. For equipment that operates on the principle of Stokes' law, the diameter of the oil droplet has a major effect on the separation and removal of the oil droplet from the water.

The capability of a given de-oiling device or system to remove and recover dispersed oil decreases as the droplet size decreases. Oil droplet size distribution is a fundamental characteristic of produced water and must be considered in designing and sizing treating systems to meet regulatory standards for effluent water compliance.

Figure 3.1 is an example of a typical histogram of an oil droplet distribution. The histogram divides the particle counts into discrete size ranges along the horizontal axis. The number and size of the ranges are determined by the equipment used to obtain the data. The height of the vertical bars corresponds to the volume percentage of oil droplets in each range. A particle distribution curve is constructed by connecting the tops of the bars at the midpoint of each size range.

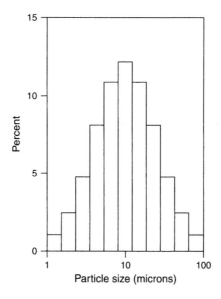

FIGURE 3.1. Histogram of oil droplet distribution.

Figure 3.2 illustrates a typical volume distribution curve. The volume percentage of the particles is equal to or smaller than each specified size that is plotted. The vertical axis scale is from 0–100% since the data are plotted cumulatively. Volume percent distribution curves are generally more directly useful for designing water de-oiling equipment.

The dispersed oil droplet size distribution may vary from point to point in a produced water system, and from one system to another. The size distribution is affected by interfacial tension, turbulence, temperature, system shearing (pumping, pressure drop across pipe fitting, etc.), and other factors. The droplet size distributions should be measured in the field when troubleshooting and/or upgrading systems, whenever possible.

In the absence of data, the generalized relationship in Figure 3.3 can be used for oil droplet size distributions. Since the distribution is linear, it places more volume in smaller-diameter droplets. However, because this straight-line relationship is a very rough estimate, field data should be used whenever possible. For produced water effluent from a three-phase separator, a maximum oil droplet diameter of 250–500 μm and an oil content of 1000–2000 mg/l can be used in the absence of field data. For first phase de-oiling equipment, an oil droplet diameter of 30 μm with inlet total oil levels less than 100 mg/l can be assumed for produced water feed to final treating equipment. Operational experience in the area may also provide reliable data from similar existing facilities that can be used to estimate inlet oil concentrations and droplet size distributions.

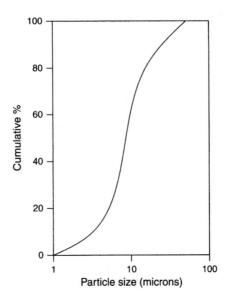

FIGURE 3.2. Typical oil volume distribution curve.

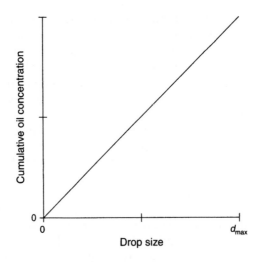

FIGURE 3.3. Droplet size distribution for design.

3.3.10 Toxicants

Produced water can exhibit toxicity to marine organisms in laboratory tests. The potential for toxic effects in the natural environment is one reason for concern about the environmental effects of produced water discharges.

The toxicity of produced water is determined by exposing groups of test organisms to a series of produced water concentrations in seawater for a fixed period of time. The cumulative effect is measured as a function of concentration. The object of the test is to observe effects of the test organisms such as mortality, reduction in rate of growth, and reduction in ability to reproduce.

Test results are expressed either as the maximum concentration of produced water that will produce no effect on the test organisms [the "No Observable Effect Concentration" (NOEC)] or as the concentration that produces a 50% effect on the test organisms.

The effect can be mortality, in which case the test result is referred to as an "LC-50," which stands for the concentration that is lethal to half of the test organisms. A numerically lower LC-50 indicates a more toxic effluent. For effects on growth, or other indicators, the test result is expressed as an "EC-50," which stands for the concentration that produces an effect on 50% of the test organisms.

Produced water toxicity varies widely. For a given effluent and test organism, higher concentrations are needed to produce observable effects during shorter exposure times. Some produced waters are essentially nontoxic, requiring concentrations in excess of 10% to produce effects in 4-day exposure tests. As a general rule, "acute" effects from exposure of 24 h or less can occur at concentrations in excess of 1%, whereas "chronic" effects can occur for longer exposures at concentrations in excess of 0.1%. The most toxic produced waters can cause acute effects during 24-h exposures at concentrations as low as 0.5%, although such effects are generally seen only at concentrations in excess of 1%. The effects at concentrations as low as 0.1% can be observed in 96-h exposure tests. A study of the acute toxicity to mysid shrimp in the Gulf of Mexico produced waters found a range of LC-50 values of 0.1–86% produced water, with an average of 19%. Seven-day exposure tests are used to measure "chronic" toxicity of produced waters for regulatory purposes in the United States. The average chronic "NOEC" for four Gulf of Mexico produced waters was found to be 1.6%.

Field studies show produced water discharged into the open ocean is diluted to concentrations of 1% or less within a few meters of the discharge pipe. Dilution to concentrations below a few tenths of a percent typically occurs within 300 ft (100 m) of the discharge pipe. Furthermore, the produced water plume occupies only a small fraction of the water column and is constantly moving due to local currents. As a result, it is highly unlikely that organisms in the marine environment will be exposed to elevated produced water concentrations for the long exposure times used in laboratory toxicity tests. The rapid initial dilution of produced water discharges and long exposure times needed to cause observable toxicity greatly reduce the potential for toxic effects on marine life from produced water discharges into the open ocean.

Proper outfall design can significantly reduce the potential for toxic effects from produced water discharges. Outfalls should be positioned such that the effluent plume does not contact the sea bottom. Bottom contact greatly reduces the rate of dilution and makes it possible for the produced water to have a direct impact on the organisms on the ocean floor. Diffusers (multi-port outfalls) may be used to increase dispersion and reduce the potential for plume contact with the sea floor.

Produced water may contain dispersed oil, dissolved oil, metals, ammonia, treating chemicals, and salts. Each of these constituents could act as a source of toxicity. Published research results indicate that organic compounds in produced water are significant factors in toxicity but not the source of toxicity in all cases. Common industry practice for water treating is to reduce the dispersed oil content of produced water effluent and as a result may not fully treat all sources of toxicity.

Techniques for produced water treatment for toxicity reduction are still under development. Novel treatment technologies may yet be applied, but good water management of existing facilities will certainly contribute to the overall control of toxicity. Water treatment chemicals, which reduce the dispersed oil content in the effluent water, can also contribute to the toxicity of such effluent. Other chemicals used to control scale, bio-fouling, and corrosion can also contribute to effluent water toxicity, but are commonly required to reduce equipment maintenance costs. As a result, optimization of the chemicals application program can help control effluent toxicity as well as reduce the overall chemical usage costs.

Produced water toxicity is regulated only in the United States, where government permit limits the toxicity or produced water that can be discharged in the Gulf of Mexico.

3.3.11 Naturally Occurring Radioactive Materials (NORM)

NORM can be transported to the surface in produced water and can be found in production wastes, equipment, and solids at production facilities. At offshore locations, dissolved NORM are discharged along with produced water. Because of concern over human exposure to environmental radiation, oil-field NORM have received regulatory attention and managing waste has become a significant cost factor for the industry.

Oil-field NORM result from the presence of uranium and thorium in hydrocarbon bearing formations. Many oil and gas bearing formations contain shales that have higher than average concentrations of uranium and thorium. These elements occur in chemical forms that are not water-soluble under reservoir conditions (U^{238} and Th^{232}). U^{238} and Th^{232} decay into different isotopes of radium (Ra^{236} and Ra^{228}). These radium isotopes further decay into the radioactive gas called radon (Rn^{232}). Both radium and radon are soluble in formation water under reservoir conditions and can be transported to the surface along with oil, gas, and produced water.

Once produced water leaves the reservoir, decreases in temperature and pressure can lead to the precipitation of NORM scale and particulates in production equipment, where it can accumulate as hard scales, sludge, or tank bottoms. Radon in produced fluids partitions into the gas phase during primary separation and enters the gas processing stream. Radon's boiling point is between that of ethane and propane, and radon is concentrated in the natural gas liquids fraction (this is generally a problem only in a gas plant fractionation section). Accumulated NORM containing solids are periodically cleaned out of vessels during maintenance and must be disposed of in a controlled fashion. Some equipment items cannot be readily decontaminated and are subject to special handling procedures. NORM that remain in solution are disposed of by whatever process is used to dispose of the produced water.

Radium and its decay products, including radon, may be found in any equipment that comes into contact with the produced water. Radium is often associated with barium scales since radium and barium are in the same chemical family. Radon and its decay products may be found in any equipment that comes into contact with natural gas or natural gas liquids.

Oil-field NORM are an environmental concern because of the potential for human exposure to ionizing radiation. The radium and radium decay products in oil-field NORM present a hazard only if taken into the body by ingestion or inhalation. The external radiation from equipment or waste containing NORM is almost never a significant concern. The discharge of radium in produced water is of concern because it may accumulate in seafood consumed by humans. Since no established safe level exists for the intake of radium, any consumption of radium in food is of potential concern. However, for the case of radium discharged in produced water, risk assessment studies show that consumption of fish caught near produced water outfalls will not pose an unacceptable human health risk, even in the worst cases.

Regulations governing NORM focus on equipment and wastes containing NORM rather than produced water. Regulations generally specify a limit on the external radiation level from wastes or equipment above which the material must be treated as NORM and cannot be released for unrestricted use without prior decontamination. Regulations also specify the maximum acceptable radium concentration in wastes and soils for unrestricted release or disposal. Existing regulations do not limit the radium concentration in offshore produced water discharges. Operators in the U.S. Gulf of Mexico are required to measure and report the radium concentrations of their effluents to the EPA.

NORM accumulations in production equipment can be controlled in some situations but cannot be eliminated entirely. Since NORM are incorporated in scale and other precipitates, reduced NORM accumulation is a benefit of a properly managed scale control program. NORM cannot be made nonradioactive. Consequently, the emphasis in NORM waste management is on identification, control, and volume reduction. NORM site remediation activities are directed at reducing the potential for human exposure to hazardous amounts of radioactive material.

3.3.12 Bacteria

Most produced waters contain bacteria but generally in small amounts. Measurement is done according to API RP 38, "Recommended Practice for Biological Analysis of Subsurface Injection Waters." The type and number of bacteria are important when selecting a biocide program. All bacteria have many strains, and some will be immune to a specific bactericide. Thus, continued testing and periodic change of chemical may be needed. The types of bacteria are:

- *Aerobic* bacteria, which require oxygen and are present in large quantities when seawater or surface water is used for water-flood injection. Chlorine, usually from a hypochlorite generator, is used for control.
- *Anaerobic* bacteria, which grow in the absence of oxygen. One strain is the sulfate reducing bacteria (SRB) that excrete sulfide ions that form hydrogen sulfide. The associated corrosion of equipment, safety hazard from H_2S, particle plugging, potential H_2S souring of a water-flood zone, and unsightly aesthetics of iron sulfide as well as sulfide smell cause these bacteria to be a major problem. A rigorous, permanent biocide program using commercial bactericides, or a chemical (glutaraldehyde, formaldehyde, or acrolein), is needed.
- *Facultative* bacteria, which can grow in an aerobic or anaerobic environment. Their presence can create conditions aiding the growth of SRB. Specialized chemical selection is needed for control.

The API test uses a standard culture media for specific bacteria. Other media have been used or different techniques have been applied to estimate the quantity of bacteria. Field tests showed the following:

- If the total bacteria count is less than 10,000/ml (and no SRB are present), bacteria shouldn't be a problem.
- If the total bacteria count is greater than 100,000/ml, plugging of filter media or formation rock is possible and biocide control should be used.
- If the SRB count is greater than 100/ml, treatment should be initiated for a critically important injection system; counts of 100–1000/ml will require some treatment to prevent injection well plugging; counts greater than 10,000/ml will require a rigorous program of biocide control.

Protected locations are preferred sites for bacteria growth. In pipelines, pigging may be useful to remove sediments that would otherwise shield bacteria from biocides. Any place where water lies stagnant offers an ideal site for the establishment of bacterial colonies. These places include the bottoms of vessels, ahead of blind flanges, beneath corrosion products in lines, and in "rat-holes" of well bores. Growth is affected by oil or water treating chemical selection because the SRB require sulfate but also need a nutrient, which can be supplied by the carbon, nitrogen, or phosphorous in chemicals. SRB counts of a test sample with the intended chemical concentration in the produced water should be done before implementing changes.

3.4 System Description

Table 3.2 lists the various methods employed in produced water treating systems and the types of equipment that employ each method. Figure 3.4 shows a typical produced water treating system configuration. Produced water will always have some form of primary treating prior to disposal. This system could take the form of a skim tank, skim vessel, CPI, cross-flow separator, or gas flotation unit. Other than the gas flotation unit, all of these devices employ gravity separation techniques. Depending upon the severity of the treating problem, secondary treatment may be required. Secondary treatment could utilize a CPI, a cross-flow separator, or a gas flotation unit. Liquid–liquid hydrocyclones are often used either in a single stage or with a downstream skim vessel or flotation unit.

TABLE 3.2
Produced water treating equipment

Method	Equipment Type	Approximate Minimum Drop Size Removal Capacities (μm)
Gravity separation	Skimmer tanks and vessels API separators Disposal piles Skim piles	100–150
Plate coalescence	Parallel plate interceptors Corrugated plate interceptors Cross-flow separators Mixed-flow separators	30–50
Enhanced coalescence	Precipitators Filters/coalesces Free-flow turbulent coalesces	10–15
Gas flotation	Dissolved gas Hydraulic dispersed gas Mechanical dispersed gas	10–20
Enhanced gravity separation	Hydrocyclones Centrifuges	15–30
Filtration	Multimedia membrane	1+

Offshore, produced water can be piped directly overboard after treating, or it can be routed through a disposal pile or a skim pile. Water from the deck drains must be treated for removal of "free" oil. This is normally done in a skim vessel called a sump tank. Water from the sump tank is either combined with the produced water or routed separately for disposal overboard.

Onshore, the water is normally re-injected in the formation or pumped into a disposal well. In the past, particularly in dry climates in countries with emerging environmental regulations, small amounts of produced water were disposed of in an evaporation pit. This practice has virtually been ceased and thus will not be discussed any further in this text.

For safety considerations, closed drains, if they exist in the process, should never be tied into atmospheric drains and should be routed to a pressure vessel prior to entering an atmospheric skim tank or pile. This should be done in a skim vessel, or cross-flow separator in a pressure vessel.

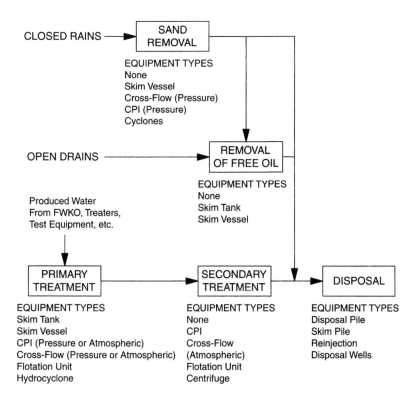

FIGURE 3.4. Typical produced water treating system.

3.5 Theory

The function of all water treating equipment is to cause the oil droplets, which are dispersed in the water continuous phase, to separate and float to the surface of the water so they can then be removed. In gravity separation units, the difference in specific gravity causes the oil to float to the surface of the water. The oil droplets are subjected to continuous dispersion and coalescence during the trip up the well bore through the surface chokes, flow lines, control valves, and process equipment. When energy is put into the system at a high rate, the drops are dispersed to smaller sizes. When the energy input rate is low, small droplets collide and join together in the coalescence process.

The three basic phenomena that are used in the design of common produced water treating equipment are gravity separation, coalescence, and flotation. Dispersion also affects the design but to an unpredictable degree. In the past filtration has been tried, but, due to high maintenance costs, has been found to be unsatisfactory.

3.5.1 Gravity Separation

Most commonly used water treating equipment depends on the forces of gravity to separate the oil droplets from the water continuous phase. The oil droplets, being lighter than the volume of water they displace, have a buoyant force exerted upon them. This is resisted by a drag force caused by their vertical movement through the water. When the two forces are equal, a constant velocity is reached, which can be computed from Stokes' law as

Field units

$$V_o = \frac{1.78 \times 10^{-6}(\Delta SG)(d_m)^2}{\mu_w},$$ (3.1a)

SI units

$$V_o = \frac{5.556 \times 10^{-7}(\Delta SG)(d_m)^2}{\mu_w},$$ (3.1b)

where

V_o = rising vertical velocity of the oil droplet relative to the water continuous phase, ft/sec (m/sec),
d_m = diameter of the oil droplet, microns (µm),
ΔSG = difference in specific gravity of oil and water relative to water,
μ_w = viscosity of the water continuous phase, cp.

Several conclusions can be drawn from this simple equation:

1. The larger the size of an oil droplet, the larger the square of its diameter and, thus, the greater its vertical velocity will be. That is, the bigger the droplet size, the less time it will take for the droplet to rise to a collection surface and thus the easier it will be to treat the water.
2. The greater the difference in density between the oil droplet and the water phase, the greater the vertical velocity will be. That is, the lighter the crude, the easier it will be to treat the water.
3. The higher the temperature, the lower the viscosity of the water and, thus, the greater the vertical velocity will be. That is, it is easier to treat the water at high temperatures than at low temperatures.

Theoretically, Stokes' law should apply to oil droplets as small as 10 µm. However, field experience indicates that 30 µm sets a reasonable lower limit on the droplet sizes that can be removed. Below this size, small pressure fluctuations, platform vibrations, etc., tend to impede the rise of the oil droplets to the coalescing surface.

3.5.2 Coalescence

The process of coalescence in water treating systems is more time-dependent than the process of dispersion. In a dispersion of two immiscible liquids, immediate coalescence seldom occurs when two droplets collide. If the droplet pair is exposed to turbulent pressure fluctuations, and the kinetic energy of the oscillations induced in the droplet pair is larger than the energy of adhesion between them, the contact will be broken before coalescence is completed. If the energy input into the system is too great, dispersion will occur, as discussed below. If there is no energy input, then the frequency of droplet collision, which is necessary to initiate coalescence, will be low, and coalescence will occur at a very low pace.

Most water treating equipment, with the exception of flotation units, and hydrocyclones, consists of vessels in which the oil droplets rise to the surface due to gravity forces. From a process standpoint, these are considered "deep bed gravity settlers." Experiments with deep bed gravity scttlcrs (rcfcr to Chapter 1 for further discussion) yield the following two qualitative conclusions:

- Doubling the residence time causes only a 10% increase in the maximum size droplet that will be grown in a gravity settler.
- The more dilute the dispersed phase (oil), the greater the residence time required to grow a given particle size will be. That is, coalescence occurs more rapidly in concentrated dispersions.

From these conclusions it shows that after an initial period of coalescence in a settler, additional retention time has a rapidly diminishing ability to cause coalescence and to capture oil droplets.

3.5.3 Dispersion

The term "dispersion" refers to the process of a discontinuous phase (oil) being split into small droplets and distributed throughout a continuous phase (water). This dispersion process occurs when a large amount of energy is input to the system in a short period of time. This energy input overcomes the natural tendency of two immiscible fluids to minimize the contacting surface area between the two fluids.

The dispersion process is diametrically opposed by coalescence, which is the process in which small droplets collide and combine into larger droplets. As the oil and water mixture flows through the piping, these two processes are simultaneously occurring. In the piping a droplet of oil splits into smaller droplets when the kinetic energy of its motion is larger than the difference in surface energy between the single droplet and the two smaller droplets formed from it. While this process is occurring, the motion of the smaller oil droplets causes coalescence to occur. Therefore, it should be possible to define statistically a maximum droplet size for a given energy input per unit mass and time at which the rate of coalescence equals the rate of dispersion.

One relationship for the maximum particle size that can exist at equilibrium was proposed by Hinze as follows:

$$d_{max} = 432\left(\frac{t_r}{\Delta P}\right)^{2/5}\left(\frac{\sigma}{\rho_w}\right)^{3/5},\tag{3.2}$$

where

d_{max} = diameter of droplet above whose size only 5% of the oil
volume is contained, μm,
σ = surface tension, dyn/cm,
ρ_w = density, g/cm^3,
ΔP = pressure drop, psi,
t_r = retention time, min.

From Equation (3.2), it can be seen that the greater the pressure drop and, thus, the shear forces that the fluid experiences in a given period of time while flowing through the treating system, the smaller the maximum oil droplet diameter will be. That is, large pressure drops that occur in small distances through chokes, orifices, throttling globe control valves, descanters, etc., result in smaller droplets.

Equation (3.2) is presented to illustrate the factors that affect drop size distribution in the system. The equation can be applied to determine a maximum droplet size that can exist downstream of a control valve or any other device that causes a large pressure drop.

The dispersion process is theoretically not instantaneous. However, it appears from field experience to occur very rapidly. For design purposes it could be assumed that whenever large pressure drops occur, all droplets larger than d_{max} will instantaneously disperse. This is, of course, a conservative approximation.

Unfortunately, Equation (3.2) cannot be used directly to predict the coalescence of droplets that occur in piping with high-pressure drops downstream of a process component in which dispersion takes place. This is because the coalescence to a new d_{max} determined in Equation (3.2) is time-dependent, and there is currently no basis to estimate the time required to grow d_{max}.

3.5.4 Flotation

Flotation is a process that involves the injection of fine gas bubbles into the water phase. The gas bubbles in the water adhere to the oil droplets. The buoyant force on the oil droplet is greatly increased by the presence of the gas bubble. Oil droplets are then removed when they rise to the water surface, where they are trapped in the resulting foam and skimmed off the surface. Experimental results show that very small oil droplets (greater than 10 μm) in a very dilute suspension

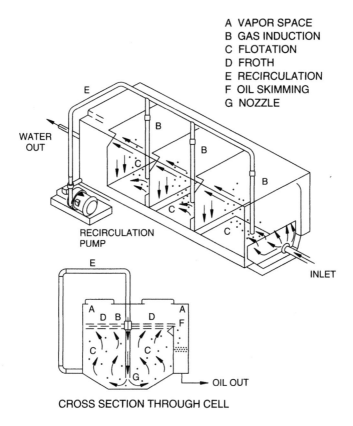

A VAPOR SPACE
B GAS INDUCTION
C FLOTATION
D FROTH
E RECIRCULATION
F OIL SKIMMING
G NOZZLE

FIGURE 3.5. Dispersed gas flotation unit with inductor.

can be removed by flotation. High percentages (90%+) of oil removal are achieved in very short times.

Figure 3.5 shows a cross section of a three-cell inductor dispersed gas flotation unit. Clean water from the effluent is pumped to a recirculation header (E) that feeds a series of venturi eductors (B). Water flowing through the eductor sucks gas from the vapor space (A) that is released at the nozzle (G) as a jet of small bubbles. The bubbles rise, causing flotation in the chamber (C), forming a froth (D) that is skimmed with a mechanical device at (F).

It would be extremely difficult to develop a precise mathematical model of the process occurring in the zones identified in this cross section. However, with the aid of some liberal assumptions, it is possible to develop a qualitative model of the efficiency of such a cell and gain an understanding of the importance of various parameters. The efficiency of a specific cell with constant geometry can be approximated through the use of Equations (3.3) through (3.5). Since these equations are presented to provide a qualitative "feel" for the effects of various parameters on flotation cell efficiency, units are not listed.

In using these equations, however, one must use parameters with consistent units.

$$E = \frac{C_i - C_o}{C_i}, \tag{3.3}$$

$$E = \frac{K}{Q_w + K}, \tag{3.4}$$

$$K = \frac{6\pi K_p r^2 h q_g}{q_w d_b}, \tag{3.5}$$

where

E = efficiency per cell,
C_i = inlet oil concentration,
C_o = outlet oil concentration,
Q_w = liquid flow rate,
K_p = mass transfer coefficient,
r = radius of mixing zone,
h = height of mixing zone,
q_g = gas flow rate,
q_w = liquid flow through the mixing zone,
d_b = diameter of gas bubble.

The following conclusions can be drawn from Equation (3.5):

1. Removal efficiency is independent of the influent oil concentration or the oil droplet size distribution.
2. Decreasing the diameter of the gas bubbles without changing the gas flow rate increases the efficiency.
3. Increasing the gas flow rate increases the efficiency.
4. Increasing the bulk flow rate decreases the efficiency.

Equation (3.5) cannot be used directly. It depends on the design details of the particular unit, which is under the control of the manufacturer, and depends on the mass flow transfer coefficient, which is a function of the composition and chemical treatment of the liquid. Most manufacturers attempt to design each cell for a typical efficiency in excess of 50%. The overall efficiency of a multiple cell flotation unit can be calculated from Equation (3.6):

$$E_t = 1 - [1 - E]^n, \tag{3.6}$$

where

E_t = overall efficiency,
n = number of stages or cells.

For an average design efficiency of 50% per stage, the following overall efficiencies may be calculated:

No. of Cells (n)	Overall Efficiency (E_t)
1	0.50
2	0.75
3	0.87
4	0.94
5	0.97

Most flotation units consist of three or four cells. Using more cells may not be cost-effective for the small performance increases shown above.

As an additional consideration, each cell must have some retention time so that the gas bubbles may have time to rise to the liquid surface. It is recommended that a minimum water retention time of 1 minute be provided in each cell.

Flotation units function best if the water flow through the unit is smooth. Therefore, it is recommended that throttling level controls be used to control the level in the upstream components of the system and in the flotation unit.

3.6 Equipment Description and Sizing

3.6.1 Skim Tanks and Skim Vessels

The simplest form of primary treating equipment is a skim (clarifier) tank or vessel; refer to Figure 3.6. These items are normally designed to provide long residence times during which coalescence and gravity separation can occur. Skim tanks can be used as atmospheric tanks,

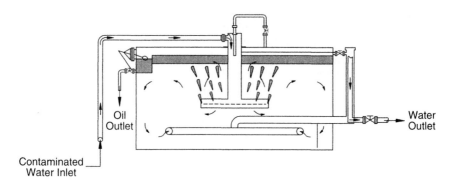

FIGURE 3.6. Schematic of a skimmer tank.

pressure vessels, and surge tanks ahead of other produced water treating equipment.

The terminology used to describe the different equipment often is a source of great confusion. A "skim (clarifier) tank" is the terminology used to describe a tank that is used to remove dispersed oil. A "settling tank," however, is the terminology used to describe tanks whose primary purpose is to remove entrained solids. On the other hand, "wash tanks" function as a free-water knockout or gunbarrel and are used when the incoming stream contains 10–90% oil. They are designed to make only a rough separation of the oil and water. The water from wash tanks is generally sent to a skim (clarifier) tank or another unit to remove the remaining oil.

If the desired outlet oil concentration is known, the theoretical dimensions of the vessel can be determined. Unlike the case of separation, with skim vessels one cannot ignore the effects of vibration, turbulence, short-circuiting, etc. American Petroleum Institute (API) Publication 421, *Management of Water Discharges: Design and Operation of Oil–Water Separators*, uses short-circuit factors as high as 1.75 and is the basis upon which many of the sizing formulas in this chapter were derived.

Configurations
Skim vessels can be either vertical or horizontal in configuration.

Vertical
In vertical skimmers the oil droplets must rise upward countercurrent to the downward flow of the water. Some vertical skimmers have inlet spreaders and outlet collectors to help even the distribution of the flow, as shown in Figure 3.7. The oil, water, and any flash gases are introduced below the oil–water interface. Small amounts of gas liberated from the water help to "float" the oil droplets. In the quiet zone between the spreader and the water collector, some coalescence can occur, and the buoyancy of the oil droplets causes them to rise counter to the water flow. Oil will be collected and skimmed off the surface.

The thickness of the oil pad depends on the relative heights of the oil weir and the water leg and on the difference in specific gravity of the two liquids. Often, an interface lever controller is used in place of the water leg.

Horizontal
In horizontal skimmers the oil droplets rise perpendicular to the flow of the water, as shown in Figure 3.8. The inlet enters in the water section so that the flashed gases may act as a dissolved gas flotation cell. The water flows horizontally for most of the length of the vessel.

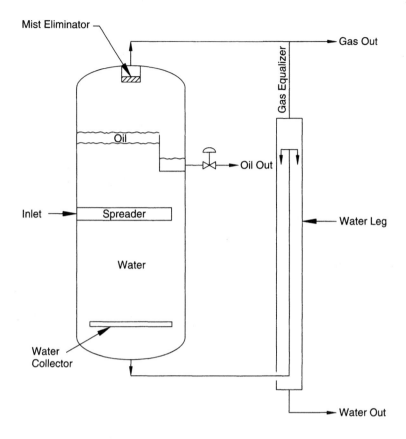

FIGURE 3.7. Schematic of a vertical skimmer vessel.

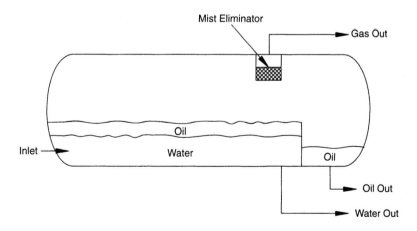

FIGURE 3.8. Schematic of a horizontal skimmer vessel.

Baffles could be installed to straighten the flow. Oil droplets coalesce in this section of the vessel and rise to the oil–water surface, where they are captured and eventually skimmed over the oil weir. The height of the oil can be controlled by interface control, by a water leg similar to that shown in Figure 3.7, or by a bucket and weir arrangement.

Horizontal vessels are more efficient at water treating because the oil droplets do not have to flow countercurrent to the water flow. However, vertical skimmers are used in instances where:

1. Sand and other solid particles must be handled. This can be done in vertical vessels with either the water outlet or a sand drain off the bottom. Experience with elaborately designed sand drains in large horizontal vessels is expensive, and they have been only marginally successful in field operations.
2. Liquid surges are expected. Vertical vessels are less susceptible to high-level shutdowns due to liquid surges. Internal waves due to surging in horizontal vessels can trigger a level float even though the volume of liquid between the normal operating level and the high-level shutdown is equal to or larger than that in a vertical vessel. This possibility can be minimized through the installation of stilling baffles in the vessel.

It should be noted that vertical vessels have some drawbacks that are not process-related and that must be considered in making a selection. For example, the relief valve and some of the controls may be difficult to service without special access platforms and ladders. The vessel may have to be removed from a skid for trucking due to height restrictions.

Pressure Versus Atmospheric Vessels

The choice of pressure versus atmospheric vessels for the skimmer tank is not determined solely by the water treating requirements. The overall needs of the system must be considered in this decision. Pressure vessels are more expensive than tanks. However, they are recommended where:

1. Potential gas blow-by through the upstream vessel dump system could create too much back-pressure in an atmospheric vent system.
2. The water must be dumped to a higher level for further treating and a pump would be needed if an atmospheric vessel were installed.

Due to the potential danger from overpressure and potential gas venting problems associated with atmospheric vessels, pressure vessels are preferred downstream of pressurized three-phase separators. However, an individual cost/benefit decision must be made for each application, taking into account all the requirements of the system.

Retention Time

Skim tanks are often used as the primary produced water treating equipment. The oil concentration of the inlet water entering the skim tank ranges from 500–10,000 mg/l. A minimum residence time of 10–30 min should be provided to assure that surges do not upset the system and to provide for some coalescence. The minimum droplet size removal is in the 100- to 300-μm range. As previously discussed, the potential benefits of providing much more residence time will probably not be cost-efficient beyond this point. Skimmers with long residence times require baffles to attempt to distribute the flow and eliminate short-circuiting. Tracer studies have shown that skimmer tanks, with carefully designed spreaders and baffles, exhibit improved flow behavior. Figure 3.9 is a schematic of a vertical skim tank with baffles.

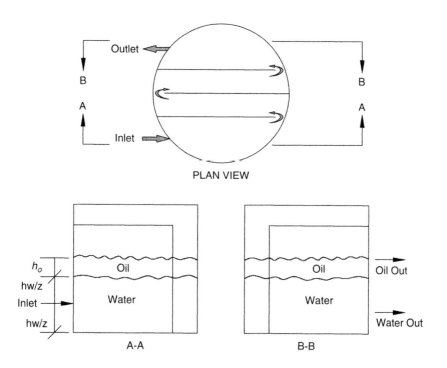

FIGURE 3.9. Schematic of a vertical skim tank with baffles.

Performance Considerations

Several factors can affect the performance of a skim tank. Some of the more important factors include:

- Carefully designed inlet and outlet distributors significantly improve the performance of a skim tank.
- Higher inlet water temperatures improve the oil removal due to a reduction in the bulk water phase viscosity.
- A short, wide, "stocky" tank is preferred over a tall, slender tank because it offers a lower downward water velocity, which aids in gravity separation.
- Unbaffled tanks are inefficient due to short-circuiting.
- Short-circuiting is reduced by installing a single vertical baffle.
- Horizontal baffles improve skim tank performance; however, to achieve maximum benefit, they should be installed as close to the horizontal as possible and caution should be used during operating maintenance not to alter the baffle arrangement.
- Often water treating chemicals, such as flocculants, are added upstream of the skim vessel. These chemicals work effectively to remove the smaller oil droplets by attaching to the oil droplets and causing them to rise to the oil–water interface in the skim vessel. However, if the chemical dosage is not carefully monitored, especially when the water rate decreases, an excess of chemical flocculants will result in a froth layer at the oil–water interface. This froth can cause the level controller to malfunction, leading to oil potentially spilling out of the vessel. Therefore, if chemical injection is used, its dosage should be carefully controlled.

Skim vessels are recommended when:

- Pressure reduction from a separator is required to protect downstream produced water treating equipment.
- Degassing water catching oil slugs or controlling surges is desired and the skim vessel is between the upstream separator and downstream produced water treating equipment.
- An existing vessel can be converted or space is available for a new vessel.
- The inlet oil concentration is high and the effluent must be reduced to 250 mg/l for the downstream equipment.
- Solid contaminants are in the inlet stream.

Skim vessels are not recommended when

- Influent oil droplet sizes are mostly below 100 μm.
- Size and weight are the primary considerations.
- Offshore structure (platform, tension leg, etc.) movement could generate waves in the vessel.
- Water temperature is very cold due to long subsea pipelines connected to other platforms.

Skimmer Sizing Equations
Horizontal Cylindrical Vessel: Half-Full

The required diameter and length of a horizontal cylinder operating 50% full of water can be determined from Stokes' law as follows assuming a 1.8 factor for turbulence and short circuiting:

Field units

$$dL_{eff} = \frac{1000 Q_w \mu_w}{(\Delta SG)(d_m)^2},$$

(3.7a)

SI units

$$dL_{eff} = 1,145,734 \frac{Q_w \mu_w}{(\Delta SG)d_m^2},$$

(3.7b)

where

d = vessel internal diameter, in. (mm),
Q_w = water flow rate, bwpd (m^3/h),
μ_w = water viscosity, cp,
d_m = oil droplet diameter, μm,
L_{eff} = effective length in which separation occurs, ft,
ΔSG = difference in specific gravity between the oil and water relative to water.

Any combination of L_{eff} and d that satisfies this equation will be sufficient to allow all oil particles of diameter d_m or larger to settle out of the water.

In addition to the settling criteria, a minimum retention time should be provided to allow coalescence. As stated earlier, increasing the retention time beyond that required for initial coalescence is not cost-effective for increasing oil droplet diameter. However, some initial retention time can be cost-effective in increasing the oil droplet size distribution. Typically, retention times vary from 10–30 min. It is recommended that a retention time of not less than 10 min be provided in skimmers that have no means of promoting coalescence. To ensure that the appropriate retention time has been provided, the following equation must also be satisfied when one selects d and L_{eff}:

Field units

$$d^2 L_{eff} = 1.4(t_r)_w Q_w, \qquad (3.8a)$$

where $(t_r)_w$ is retention time, min,

SI units

$$d^2 L_{eff} = 4.2 \times 10^4 (t_r) Q_w, \qquad (3.8b)$$

where $(t_r)_w$ is retention time, min.

The choice of correct diameter and length can be obtained by selecting various values for d and L_{eff} for both Equations (3.7a)–(3.8b). For each d, the larger L_{eff} must be used to satisfy both equations.

The relationship between the L_{eff} and the seam-to-seam length of a skimmer depends on the physical design of the skimmer internals. Some approximations of the seam-to-seam length may be made based on experience as follows:

$$L_{ss} = \frac{4}{3} L_{eff}, \qquad (3.9)$$

where L_{ss} is seam-to-seam length, in m (ft).

This approximation must be limited in some cases, such as vessels with large diameters. Therefore, the L_{eff} should be calculated using Equation (3.9) but must be equal to or greater than the values calculated using the following equations:

Field units

$$L_{ss} = L_{eff} + 2.5. \qquad (3.10a)$$

SI units

$$L_{ss} = L_{eff} + 0.76. \qquad (3.10b)$$

Field units

$$L_{ss} = L_{eff} + \frac{d}{24}. \qquad (3.11a)$$

SI units

$$L_{ss} = L_{eff} + \frac{d}{2000}. \qquad (3.11b)$$

Equations (3.10a) and (3.10b) will govern only when the calculated L_{eff} is less than 7.5 ft (2.3 m). The justification for this limit is that some minimum vessel length is always required for oil and water collection before dumping. Equations (3.11a) and (3.11b) governs when one half the diameter in feet exceeds one third of the calculated L_{eff}. This constraint ensures that even flow distribution can be achieved in short vessels with large diameters.

Horizontal Rectangular Cross-Section Skimmer
Similarly, the required width and length of a horizontal tank of rectangular cross section can be determined from Stokes' law using an efficiency factor of 1.9 for turbulence and short-circuiting:

Field units

$$WL_{eff} = 70 \frac{Q_w \mu_w}{(\Delta SG)(d_m)^2},$$ (3.12a)

SI units

$$WL_{eff} = 950 \frac{Q_w \mu_w}{(\Delta SG)(d_m)^2},$$ (3.12b)

where

W = width, ft (m),
L_{eff} = effective length in which separation occurs, ft (m).

Typically, the height of the water flow is limited to less than one-half the width to assure good flow distribution. With this assumption, the following equation can be derived to ensure that sufficient retention time is provided:

Field units

$$W^2 L_{eff} = 0.008(t_r)_w Q_w.$$ (3.13a)

SI units

$$W^2 L_{eff} = 0.0333(t_r)_w Q_w.$$ (3.13b)

The choice of W and L that satisfies both requirements can be obtained graphically. The height of water flow, H, is set equal to $0.5W$.

As with horizontal cylindrical skimmers, the relationship between L_{eff} and L_{ss} is dependent on the internal design. Approximations of the L_{ss} of rectangular skimmers may be made using the following:

$$L_{ss} = L_{eff} + \frac{W}{20}.$$ (3.14)

As before, the L_{ss} should be the largest of Equations (3.9)–(3.10b) and (3.14).

Vertical Cylindrical Skimmer
One can determine the required diameter of a vertical cylindrical tank by setting the oil rising velocity equal to the average water velocity as follows:

Field units

$$d^2 = 6691 F \frac{Q_w \mu}{(\Delta SG) d_m^2},$$ (3.15a)

SI units

$$d^2 = 6.365 \times 10^8 F \frac{Q_w \mu}{(\Delta SG) d_m^2},$$ (3.15b)

where F is a factor that accounts for turbulence and short-circuiting.

For small-diameter skimmers [48 in. or less (1.2 m or less)], the short-circuiting factor should be equal to 1.0. Skimmers with diameters greater than 48 in. (1.2 m) require a value for F. Inlet and outlet spreaders and baffles affect the flow distribution in large skimmers; therefore, they affect the value of F. It is recommended that for large-diameter skimmers, F should be set equal to $d/48$. Substituting this into Equations (3.15a) and (3.15b) gives the following:

Field units

$$d = 140 \frac{Q_w \mu_w}{(\Delta SG) d_m^2}.$$ (3.16a)

SI units

$$d = 5.3 \times 10^9 \frac{Q_w \mu_w}{(\Delta SG) d_m^2}.$$ (3.16b)

The height of the water column in a vertical skimmer can be determined for a selected d from retention time requirements:

Field units

$$H = 0.7 \frac{(t_r)_w Q_w}{d^2},$$ (3.17a)

SI units

$$H = 21{,}218 \frac{(t_r)_w Q_w}{d^2},$$ (3.17b)

where H is the height of the water, in ft (m).

The height of the oil pad in both vertical and horizontal skimmers typically ranges from 2–6 in. (50–150 mm). It is important to remember that the purpose of a water skimmer is to remove oil from water and produce as clean a water stream as possible. The quality of the skimmed oil from a skimmer is a secondary consideration. In fact, skimmed oil streams typically contain 20–50% water. The objective is to maximize the water treating ability of the skimmer. Maintaining a minimum oil pad thickness accomplishes this objective.

3.6.2 Coalescers

Several different types of devices have been developed to promote the coalescence of small dispersed oil droplets. These devices use gravity separation similar to skimmers but also induce coalescence to improve the separation. Thus, these devices can either match the performance of a skimmer in less space or offer improved performance in the same space.

Plate Coalescers

The use of flow through parallel plates to help gravity separation in skim tanks was pioneered in the late 1950s as a method of modifying existing refinery horizontal, rectangular cross-section separators to treat oil droplets less than 150 μm in diameter. Various configurations of plate coalescers have been devised. These are commonly called parallel plate interceptors (PPI), corrugated plate interceptors (CPI), or cross-flow separators. All of these depend on gravity separation to allow the oil droplets to rise to a plate surface where coalescence and capture occur. Plate coalescers overcome the size and weight disadvantage of skim tanks by enhancing coalescence of the oil droplets, thereby substantially increasing their rise velocities. Consequently, plate coalescers require smaller cross-sectional areas, thus providing space and weight gains over skim tanks.

As shown in Figure 3.10, flow is split between a number of parallel plates spaced 0.5–2 in. (1.2–5 cm) apart. To facilitate capture of the oil droplets, the plates are inclined to the horizontal, which promotes

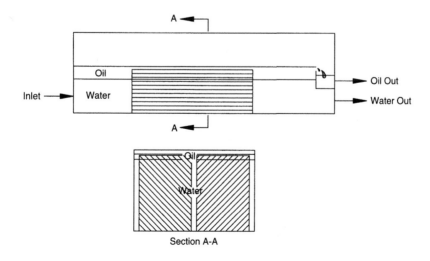

FIGURE 3.10. Schematic of a parallel plate interceptor.

oil droplet coalescence into films, and to guide the oil to the top for entrapment into channels, thereby preventing remixing with the water. The plates provide a surface for the oil droplets to collect and for solid particles to settle.

Figure 3.11 shows that an oil droplet entering the space between the plates will rise in accordance with Stokes' law. At the same time, the oil droplet will have a forward velocity equal to the bulk water velocity. By solving for the vertical velocity needed by a particle entering at the base of the flow to reach the coalescing plate at the top of the flow, the resulting droplet diameter can be determined.

It is important to note that Stokes' law should apply to oil droplets as small in diameter as 1–10 μm. However, field experience indicates that 30 μm sets a reasonable lower limit on the droplet sizes that can be removed. Below this size small pressure fluctuations, platform vibration, etc., tend to impede the rise of the droplets to the coalescing surface.

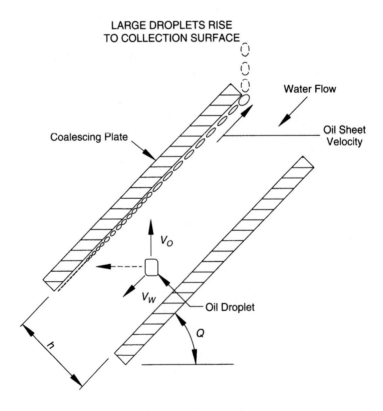

FIGURE 3.11. Cross section showing plate coalescer operation.

Parallel Plate Interceptor (PPI)

The first form of a plate coalescer was the PPI, as shown in Figure 3.10. This involved installing a series of plates parallel to the longitudinal axis of an API separator (a horizontal, rectangular cross-section skimmer). The plates form a "V" when viewed along the axis of flow so that the oil sheet migrates up the underside of the coalescing plate and to the sides. Sediments migrate toward the middle and down to the bottom of the separator, where they are removed. The interplate spacing can be small, which would allow packing more plates inside a vessel, which would in turn maximize the area for oil droplets to coalesce. However, this spacing would increase the probability of plugging the interspaces with solids. As a compromise, a distance of 3/4 in. is typically used. The angle of inclination for the plates is generally established at 45°.

Corrugated Plate Interceptor (CPI)

The most common form of PPI used in production operations is the CPI. This is a refinement of the PPI in that it takes up less plan area for the same particle size removal, it makes sediment handling easier, and it has the added benefit of being cheaper than a PPI.

Figure 3.12 shows the flow pattern of a typical downflow CPI design. Water enters the inlet nozzle (1), where solids flow downward and settle in the primary collection box (2). Water and oil flow up and through a perforated distribution baffle plate (3). The CPI pack (4) receives oily water. Oil rises out of the flow path to the underside of the ridge and coalesces into a film moving upward opposite the bulk water flow. A thick layer of oil is allowed to collect until it flows over an adjustable weir (5) into an oil collection box for removal. Light solids and sludge separation is simultaneously accomplished and falls to the lower plate surface along the gutters and collects at the bottom (6), where it is removed. After exiting the CPI pack, the water moves upward and flows over an adjustable weir (7) into the water removal collection box. A secondary oil removal outlet (8) is located above the water outlet. A gasketed cover (9) allows for gas blanket operation. It is also supplied with an adequately sized vent nozzle (10).

In CPIs the parallel plates are corrugated (like roofing material), and the axes of the corrugations are parallel to the direction of flow. Figure 3.13 shows a typical CPI pack. The plate pack is inclined at an angle of 45° and the bulk water flow is forced downward. The oil sheet rises upward counter to the water flow and is concentrated in the top of each corrugation. When the oil reaches the end of the plate pack, it is collected in a channel and brought to the oil–water interface.

In areas where sand or sediment production is anticipated, the sand should be removed prior to flowing through a standard CPI.

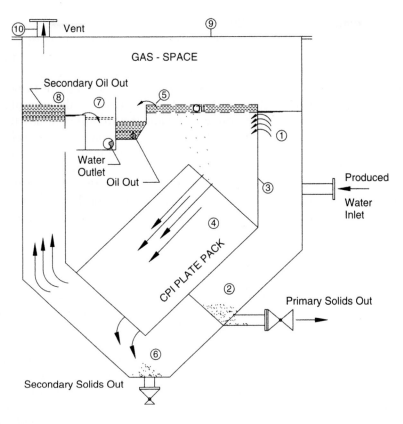

FIGURE 3.12. Schematic showing flow pattern of a typical down-flow CPI design.

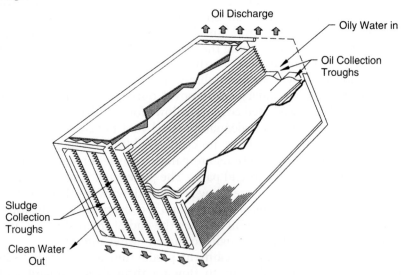

FIGURE 3.13. CPI plate pack.

Because of the required laminar flow regime, all plate coalescers are efficient sand settling devices.

Experience has shown that oil wet sand may adhere to a 45° slope. Therefore, the sand may adhere to and clog the plates. In addition, the sand collection channels installed at the end of the plate pack cause turbulence that affects the treating process and are themselves subject to sand plugging. To eliminate the above problems, an "up-flow" CPI unit employing corrugated plates, spaced a minimum of 1 in. (2.5 cm) apart with a 60° angle of inclination, may be used. Water jets for sand removal should also be installed. Figure 3.14 is a schematic showing the flow pattern of a typical up-flow CPI design. Figure 3.15 compares the flow pattern of an up-flow and down-flow CPI pack.

The main components of a CPI plate separator are:

- separator basin,
- CPI plate pack,
- oil and effluent weir,

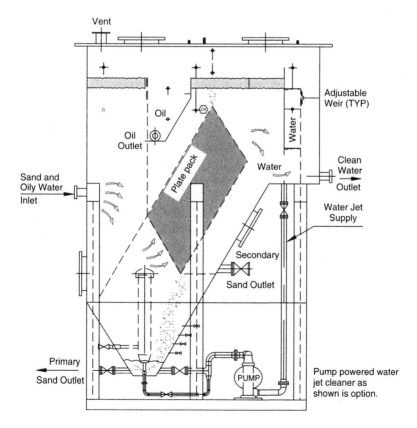

FIGURE 3.14. Schematic showing flow pattern of a typical up-flow CPI design.

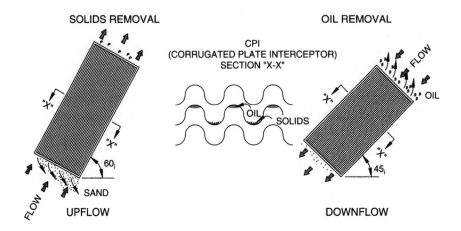

FIGURE 3.15. Up-flow versus down-flow flow pattern.

- basin cover,
- solids hopper, and
- inlet and outlet nozzles.

The *separator basin* and its internals are generally made of carbon steel plate with at least a 3/16-in. thickness. The basin edges are welded. All carbon steel external and internal surfaces are blast cleaned and painted with epoxy paint.

The *CPI plate packs* are constructed of chlorinated polyvinylchloride (CPVC), polyvinylchloride (PVC), polypropylene (PP), fiberglass reinforced polyester, carbon steel, galvanized steel, or various grades of stainless steel. Stainless steel plate packs can be used up to temperatures as high as 350 °F (125 °C), whereas the polymer plates are limited to about 140 °F (55 °C). The plate pack usually has a 316 SS frame for robustness and easy removal during maintenance. Polypropylene plates have an inherently oleophilic property that attracts oil, thus promoting coalescence. Polypropylene also repels water, which aids the downward flow of sludge, thus reducing chances of sludge fouling.

The *oil weir* is a bucket type and is made of carbon or stainless steel. The *effluent weir* is a plate type and is its height is adjustable.

The *basin cover* is normally made of carbon steel, heavy-duty galvanized steel, or lightweight fiberglass reinforced plastic (FRP) with 3/16-in. thickness.

The *solids hopper* may be conical or dish-shaped for cylindrical separators, or shaped like an inverted pyramid for rectangular separators.

The vessel should be leak tested prior to coating. The assembled package should be dry function tested to ensure proper operation. Any plastic piping should also be hydrotested.

Cross-Flow Devices

Equipment manufacturers have modified the CPI configuration for horizontal water flow perpendicular to the axis of the corrugations in the plates, as shown in Figure 3.16. This modification allows the plates to be put on a steeper angle to facilitate sediment removal and to enable the plate pack to be more conveniently packaged in a pressure vessel. The latter benefit may be required if gas blow-by through an upstream dump valve could cause relief problems with an atmospheric tank.

Cross-flow devices can be constructed in either horizontal or vertical pressure vessels. The horizontal vessels require less internal baffling, as the ends of almost every plate conduct the oil directly to the oil–water interface and the sediments to the sediment area below the water flow area. However, as shown in Figure 3.17, the pack is long and narrow and, therefore, it requires an elaborate spreader and collection device to force the water to travel across the plate pack in plug flow. The inlet oil droplets may shear in the spreader, which would make separation more difficult. This configuration would be preferred when a pressure vessel in a high-pressure system is needed.

Vertical units, although requiring collection channels on one end to enable the oil to rise to the oil–water interface and on the other end to allow the sand to settle to the bottom, can be designed for more efficient sand removal.

The cross-flow device may be installed in an atmospheric vessel, as shown in Figure 3.18, or in a vertical pressure vessel.

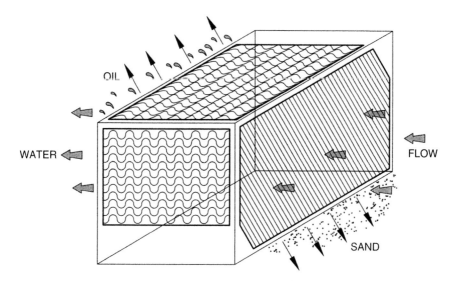

FIGURE 3.16. Schematic showing flow pattern of cross-flow plate pack.

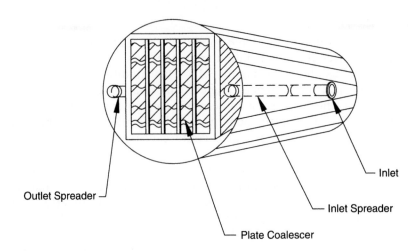

FIGURE 3.17. Schematic showing cross-flow device installed in a horizontal pressure vessel.

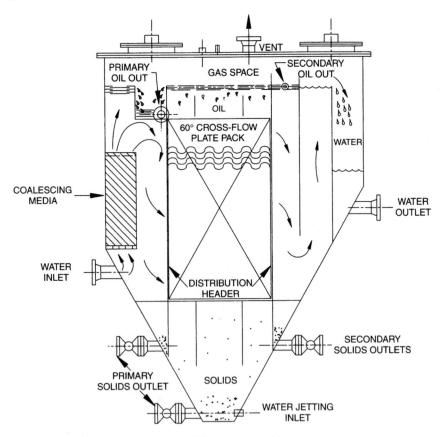

FIGURE 3.18. Schematic showing cross-flow device installed in an atmospheric vessel.

CPI separators are generally cheaper and more efficient at oil removal than cross-flow separators. However, cross-flow separators should be considered where a pressure vessel is preferred or where high sand production is expected and the sand is not removed upstream of the water treating equipment.

Performance Considerations

Flow direction considerations include:

- *Downflow.* For efficient oil removal the downflow configuration is preferred. In this case the plate pack is inclined at a 45° angle, provided the solids content is not significant.
- *Upflow.* If the production stream contains a significant amount of solid particles, upflow CPIs with the pack inclined at 60° to the horizontal are preferred. The higher plate slope provides about 25% greater runoff force and a 30% lower erosion rate than the industry-standard 45° plate slope.
- *Cross flow.* Cross flow should be considered where the use of a pressure vessel is preferred and solids and oil removal is desired.

Plate separators generally exhibit the following advantages:

- They require very little maintenance. Coalescing packs can be easily removed as complete modules for inspection and cleaning, if necessary.
- They have smaller size and weight requirements than skim vessels because of the effect of the closely spaced inclined plates.
- They can accept fairly high concentrations of oil or solids in the inlet feed. The inlet oil influent can be as high as 3000 mg/l.
- They can separate oil droplets down to about 30 μm.
- They have a sand removal ratio of 10:1; that is, if a CPI unit captures 50-μm oil droplets, it will also capture solid particles as low as 5 μm.
- They are totally enclosed, thereby eliminating vapor losses and reducing fire hazards.
- CPIs are more efficient at oil removal than cross-flow separators are.
- They are simple and inexpensive in comparison to some of the other types of produced water treating devices, for example, flotation units.

- They have no moving parts and do not require power.
- They are easy to cover, due to their small size, and retain hydrocarbon vapors.
- They are easy to install in a pressure vessel, which helps to retain hydrocarbon vapors and protect against overpressure due to failure of an upstream level control valve.

The disadvantages of plate separators include:

- They are not effective for streams with slugs of oil.
- They cannot effectively handle large amounts of solids and emulsified streams.

Plate separators are recommended when:

- Water flow rate is steady or feed is from a pump.
- Size and weight are not constraints.
- Utilities and equipment are available to periodically clean the plate packs.
- Influent oil content is high and oil concentration must be reduced to 150 mg/l for effective second-stage treating in a downstream unit.
- Solid contaminants are not significant in the waste stream, and sand content is less than 110 ppm.

Plate separators are not recommended when:

- Influent droplet sizes are mostly below 30 µm.
- Size and weight are the primary considerations.
- Sand particle diameters are less than 25 µm, and solids removal is a primary objective.

Selection Criteria

Plate separators are effective to approximately 30 µm. Vendor-supplied nomographs can be used to estimate the performance of CPIs. Figure 3.19 presents a relationship among the liquid inflow temperature, particle size removed, differential specific gravity of the oil and water, and capacity for downflow oil removal with a 45° plate angle.

For example, produced water flowing at a rate of 150 gpm (5143 bbl/day) per CPI pack with 3/4-in. spacing, a differential specific gravity of 0.1, and a flowing temperature of 68°F will remove a particle of about 60 µm. Similarly, Figure 3.20 is a nomograph for downflow oil removal with a 60° plate angle, and Figure 3.21 has the performance relationship for cross-flow oil removal.

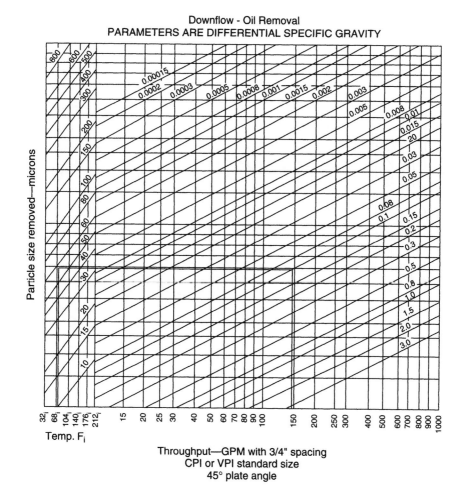

Downflow - Oil Removal
PARAMETERS ARE DIFFERENTIAL SPECIFIC GRAVITY

FIGURE 3.19. Nomograph for downflow 45° CPI.

Coalescer Sizing Equations

The general sizing equation for a plate coalescer with flow either parallel to or perpendicular to the slope of the plates for droplet size removal is

Field units

$$HWL = \frac{4.8 Q_w h \mu_w}{\cos \theta (d_m)^2 (\Delta SG)},$$

(3.18a)

SI units

$$HWL = \frac{0.794 Q_w h \mu_w}{\cos \theta (d_m)^2 (\Delta SG)},$$

(3.18b)

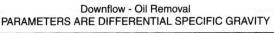

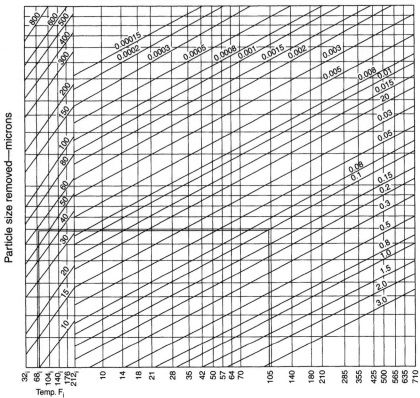

FIGURE 3.20. Nomograph for downflow 60° CPI.

where

d_m = design oil droplet diameter, μm,

Q_w = bulk water flow rate, bwpd (m³/h),

h = perpendicular distance between plates, in. (mm),

μ_w = viscosity of the water, cp,

θ = angle of the plate with the horizontal,

H, W = height and width of the plate section perpendicular to the axis of water flow, ft (m),

L = length of plate section parallel to the axis of water flow, ft (m),

ΔSG = difference in specific gravity between the oil and water relative to water.

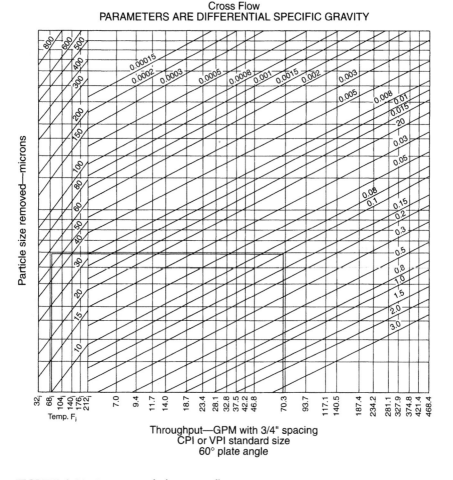

FIGURE 3.21. Nomograph for cross-flow 60° CPI.

Experiments have indicated that the Reynolds number for the flow regime cannot exceed 1600 with four times the hydraulic radius as the characteristic dimension. Based on this correlation, the minimum H times W for a given Q_w can be determined from the following which assume a surge factor of 2:

Field units

$$HW = 14 \times 10^4 \frac{Q_w h (SG)_w}{\mu_w}. \tag{3.19a}$$

SI units

$$HW = 8.0 \times 10^{-4} \frac{Q_w h (SG)_w}{\mu_w}. \tag{3.19b}$$

CPI Sizing

For CPIs, plate packs come in standard sizes with $H = 3.25$ ft (1 m), $W = 3.25$ ft (1 m), $L = 5.75$ ft (1.75 m), $h = 0.69$ in., and $\theta = 45°$. The size of the CPIs is determined by the number of standard plate packs installed. To arrive at the number of packs needed, the following equation is used:

Field units

$$\text{Number of packs} = 0.077 \frac{Q_w \mu}{(\Delta SG)(d_m)^2}. \qquad (3.20a)$$

SI units

$$\text{Number of packs} = 11.67 \frac{Q_w \mu}{(\Delta SG)(d_m)^2}. \qquad (3.20b)$$

To ensure that the Reynolds number limitation is met, the flow through each pack should be limited to approximately 20,000 bwpd.

It is possible to specify a 60° angle of inclination to help alleviate the solids plugging problem inherent in CPIs. This requires a 40% increase in the number of packs according to the following equation:

Field units

$$\text{Number of packs} = 0.11 \frac{Q_w \mu}{(\Delta SG)(d_m)^2}. \qquad (3.21a)$$

SI units

$$\text{Number of packs} = 16.68 \frac{Q_w \mu}{(\Delta SG)(d_m)^2}. \qquad (3.21b)$$

Cross-Flow Device Sizing

Cross-flow devices obey the same general sizing equations as plate coalescers. Although some manufacturers claim greater efficiency than CPIs, the reason for this is not apparent from theory, laboratory, or field tests; as a result, verification is unavailable. If the height and width of these cross-flow packs are known, Equations (3.18a) and (3.18b) can be used directly. It may be necessary to include an efficiency term, normally 0.75, in the denominator on the right side of Equations (3.18a) and (3.18b) if the dimensions of H or W are large and a spreader is needed.

Both horizontal and vertical cross-flow separators require spreaders and collectors to uniformly distribute the water flow among the plates. For this reason, the following equation has been developed assuming a 75% spreader efficiency term:

Field units

$$HWL = \frac{6.4 Q_w h \mu_w}{\cos\theta (\Delta SG)(d_m)^2}.$$ (3.22a)

SI units

$$HWL = \frac{1.0 Q_w h \mu_w}{\cos\theta (\Delta SG)(d_m)^2}.$$ (3.22b)

Example 3.1: Determining the dispersed oil content in the effluent water from a CPI plate separator

Given:
Feed water flow rate = 25,000 bbl/day at 125 °F,
Feed water specific gravity = 1.06 at 125 °F,
Feed water viscosity = 0.65 cp at 125 °F,
Dispersed oil concentration = 650 mg/l,
Dissolved oil concentration = 10 mg/l,
Total oil and grease = 660 mg/l.
The dispersed oil droplet size distribution in feed water is as follows:

Microns	<40	40–60	60–80	80–100	100–120	>120
Vol%	9	14	30	35	10	2

A vendor has quoted that one of its standard plate packs would be capable of reducing the total oil and grease content of the effluent water to less than 200 mg/l. The vendor's standard plate pack has the following geometric specification:

$H = 3.25$ ft, $W = 3.25$ ft, $L = 5.75$ ft, $h = 0.69$ in., $\theta = 45°$.

Calculate the total oil and grease content in effluent water from the plate pack to check the vendor's quoted performance.

Solution:
In order to calculate the total oil and grease in the effluent water, we must first determine the smallest oil droplet size that can be removed in the vendor's standard plate pack at the design conditions given. Equations (3.20a) and (3.20b) were derived for the given plate pack geometric configuration assumed in this example calculation.

Rearranging Equations (3.20a) and (3.20b) to solve for the minimum oil droplet size (d_m), we have the following results:

$$d_m = \sqrt{0.077 \frac{Q_w\mu}{\Delta SG(\text{No. of packs at } 45°)}} = \sqrt{0.077 \frac{(25,000)(0.65)}{(1.06 - 0.75)(1)}}$$

$$= 63.5 \text{ μm}.$$

The volume percent of the dispersed oil removed by the plate pack is determined by summing the volume percents of dispersed oil droplets contained in the feed water that are greater than or equal to 63.5 μm (see dispersed oil droplets size distribution data given). Therefore,

$$\text{Vol.\% removed} = \left(\frac{80 - 63.5}{80 - 60}\right)(30) + (35) + (10) + (2) = 71.75\%$$

Calculating the dispersed oil content in the effluent water from the plate pack (C_{out}):

$$C_{out} = (650)(100 - 71.75\%) = 183.6 \text{ mg/l}.$$

Since the plate pack does not remove any of the dissolved oil, the total oil and grease content in the effluent water from the plate pack is equal to 183.6 + 10 mg/l, or 193.6 mg/l. Therefore, the vendor's quoted performance looks to be correct.

Oil/Water/Sediment Coalescing Separators

The oil/water/sediment coalescing separator is an enhancement of the cross-flow configuration in that it utilizes a two-step process to separate small oil droplets and solids from the wellstream. The coalescing packs used are cross-flow in design rather than down-flow or up-flow. The units can be configured in either an atmospheric pressure tank (Figure 3.22) or a vertical pressure vessel (as shown in Figure 3.23). Both configurations use an inlet flow distributer/coalescer pack and a cross-flow plate pack.

The inlet flow distributer/coalscer pack evenly spreads the inlet flow over the full height and width of the separator pack. Flow through this pack is mildly turbulent, thus creating opportunities for the oil droplets to coalesce into larger ones.

The cross-flow plate pack receives flow from the distributer/coalescer pack. It consists of mutually supportive, inclined plates oriented in a hexagonal configuration. Laminar flow is established and maintained as water flows in a sinusoidal path across the pack from the inlet to the outlet. Oil rises into the top of hexagons and then along the plate's surface to the oil layer that is established at the top of the pack. The sludge slides down the plates and drops into a discreet sludge hopper in the bottom of the separator.

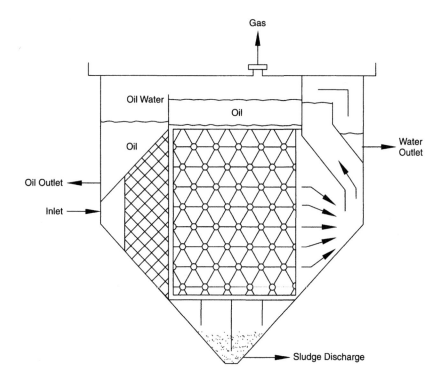

FIGURE 3.22. Schematic of an oil/water/sediment coalescing tank.

Standard spacing of cross-flow plate packs is 0.80 in., with optional available spacing of either 0.46 or 1.33 in. The pack is inclined 60° to lessen plugging. More coalescing sites are offered to the dispersed oil droplets due to the hexagonal pattern of the pack.

Coalescing pack materials include polypropylene, polyvinyl chloride, stainless steel, and carbon steel. Due to its oleophilic nature (enhances oil removal capabilities and resists plugging and fouling of the pack), polypropylene packs are commonly used up to 150 °F (55 °C). Above this temperature, the polypropylene loses pack integrity and chemical degradation begins. Stainless steel and carbon steels are used in temperatures above 150 °F (55 °C) and environments that contain large amounts of aromatic hydrocarbons.

Oil/Water/Sediment Sizing
The geometry of plate spacing and length can be analyzed for this configuration using Equations (3.18a) and (3.18b). Ten and the techniques previously discussed.

Performance Considerations
The oil/water/sediment coalescing separator exhibits the same advantages and disadvantages as plate separators. The one additional

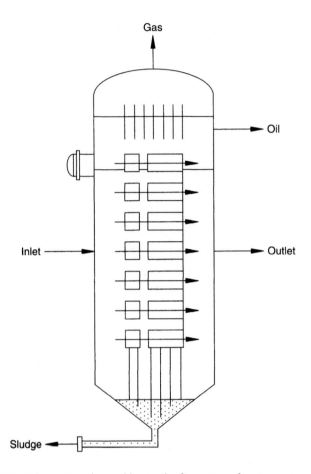

FIGURE 3.23. Schematic of an oil/water/sediment coalescing pressure vessel.

improvement is that the minimum oil droplet size that can be removed is 20 μm.

3.6.3 Skimmer/Coalescers

Several designs that are marketed for improving oil–water separation rely on installing coalescing plates or packs within horizontal skimmers or free-water knockouts to encourage coalescence and capture of small oil droplets within the water continuous phase. Coalescers act to accumulate oil on a preferentially oil wet surface where small droplets can accumulate. These larger oil droplets can be either collected directly from the oil wet surface or stripped from the oil wet surface and separated from the water phase using some type of gravity-based equipment.

Coalescing equipment may either be housed in a separate vessel or, more commonly, installed in a coalescing pack contained in a gravity vessel. Figure 3.24 shows a schematic of a horizontal FWKO with coalescing pack. The plates in a CPI or cross-flow vessel may be fabricated of an oleophilic (oil wetting) material and thereby serve as both a gravity separation device and a coalescing device. Figure 3.25 is a cross section of structured packing serving as a coalescer.

The oil wetting surface may also occur in the form of a fibrous pack or as a collection of granules. Coalescers of this type resemble filters but serve to "grow" rather than to capture oil droplets.

Matrix Type
Mats of fibers have the advantage of large surface areas and easy fabrication. Oleophilic materials are spun into thin fibers, and the fibers are collected into a pack, across which the oily water flows. Oil droplets stick to the fibers and coalesce. Figure 3.26 illustrates the coalescence process on a fibrous mat. The coalesced droplets can easily be collected after they emerge from the mat using a gravity-based separator. Figure 3.27 is an example of such an arrangement.

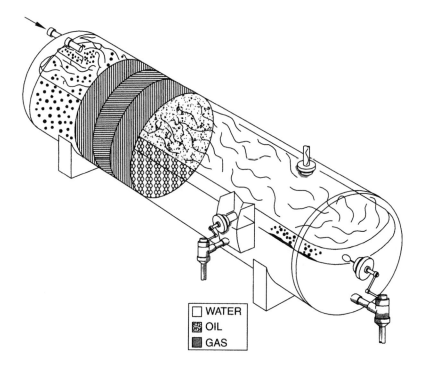

| □ WATER |
| ▦ OIL |
| ■ GAS |

FIGURE 3.24. Schematic of an FWKO with a coalescing pack.

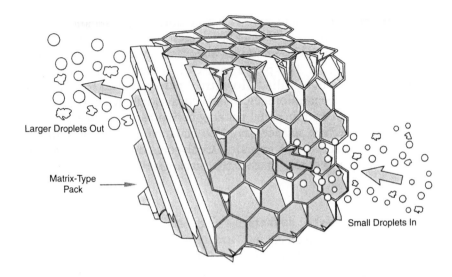

Larger Droplets Out

Matrix-Type Pack →

Small Droplets In

FIGURE 3.25. Structured packing serving as a coalescer.

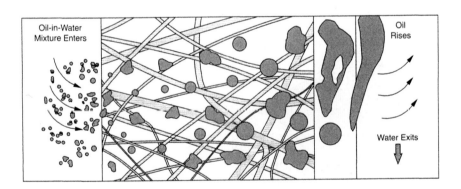

Oil-in-Water Mixture Enters

Oil Rises

Water Exits

FIGURE 3.26. Oil coalescence on a fibrous mat.

Loose Media

Oleophilic material can also be fabricated into loose media, and the media collected in a vessel. If the material is fabricated in a granular form and assembled into a deep bed gravity settler, the deep bed filter can also perform a coalescing function.

The geometry of plate spacing and length can be analyzed for each of these designs using Equations (3.18a) and (3.18b) and the techniques previously discussed. The packs cover the entire inside diameter of the vessel unless sand removal internals are required. Pack lengths range from 2–9 ft, depending on the service.

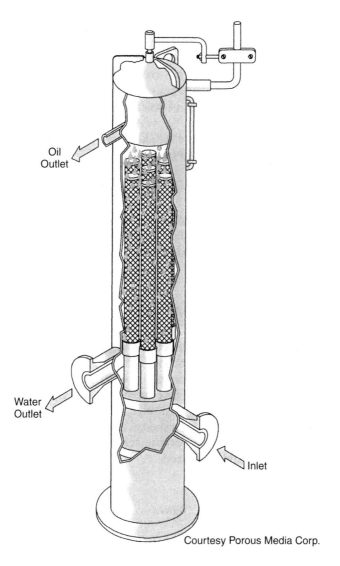

Oil
Outlet

Water
Outlet

Inlet

Courtesy Porous Media Corp.

FIGURE 3.27. Collection of oil from a matrix separator.

Performance Considerations

Coalescers are used to improve the performance of other gravity-based separation equipment. They are specified by the equipment vendor as an integral part of the water treating system, or they may be added as a retrofit to improve the performance of an existing system. Coalescers are particularly useful when the oil droplet size in the incoming water is small as a result of excess shearing in upstream piping or valves.

Coalescers can be used when:

- An existing low-pressure separator, skimmer, or plate separator can be retrofitted with a coalescing section.
- The coalescing section is accessible for cleaning or replacement.
- The inlet oil droplet size is less than 50 μm and larger droplets are desired.
- Coalescers can also serve as skimmers (the limitations listed for skimmers are applicable).

Coalescers should not be used when:

- Inlet droplet sizes are less than 10 μm.
- Inlet droplet sizes are greater than 100 μm.
- Size and weight are primary considerations.

3.6.4 Precipitators/Coalescing Filters

Precipitators are obsolete and would not be used in a new installation. In the past, it was common to direct the water to be treated through a bed of excelsior (straw) or another similar medium, as shown in Figure 3.28, to aid in the coalescing of oil droplets. However, the coalescing medium has a tendency to clog. Many of these devices in oil-field service have the medium removed. In such a case they actually act like a vertical skimmer since the oil droplets must flow counter-current to the downward flow of the water through the area where the medium was originally located.

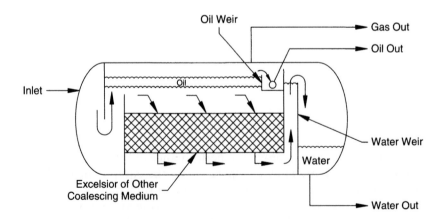

FIGURE 3.28. Schematic of a precipitator.

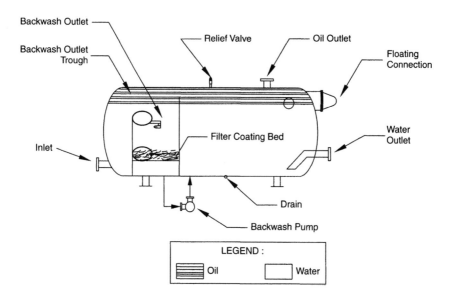

FIGURE 3.29. Schematic of a coalescer.

Coalescers, as shown in Figure 3.29, are similar in design to a precipitator except that they usually employ a larger gravity separation section than a precipitator and utilize a back-washable filter bed for coalescing and some sediment removal. The filter media are designed for automatic backwash cycles. They are extremely efficient at water cleaning, but clog easily with oil and are difficult to back-wash. The backwash fluid must be disposed of, which leads to further complications.

Some operators have had success with filters employing sand and other filter media in onshore operations where the backwash fluid can be routed to large settling tanks, and where the water has already been treated to 25–75-mg/l oil. Applications of this type are typical when the produced water will be re-injected as for a water flood.

3.6.5 Free-Flow Turbulent Coalescers

The plate coalescing devices discussed above use gravity separation followed by coalescence to treat water. Plate coalescers have the disadvantage of requiring laminar flow and closely spaced plates in order to capture the small oil droplets and keep them from stripping the coalesced sheet. They are thus susceptible to plugging with solids.

Free-flow turbulent coalescers are a type of device that is installed inside or just upstream of any skim tank or coalescer to promote coalescence. These devices had been marketed and sold under the trade name SP Packs. They are no longer available for sale but the concept can still be employed in water treating system design. As shown in Figure 3.30, SP Packs force the water flow to follow a serpentine pipe-like path sized to create turbulence of sufficient magnitude to promote coalescence, but not so great as to shear the oil droplets below a specified size. SP Packs are less susceptible to plugging since they require turbulent flow (high Reynolds numbers), have no closely spaced passages, and have a pipe path similar in size to the inlet piping.

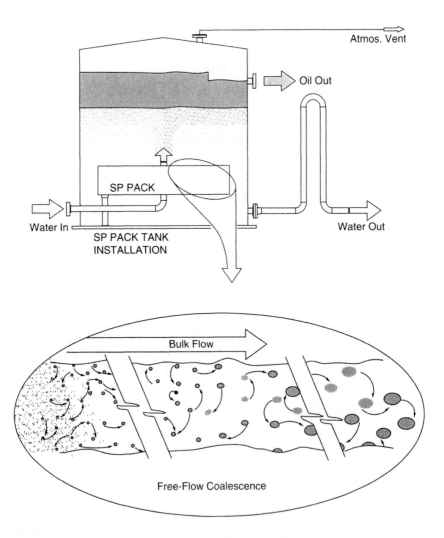

FIGURE 3.30. Principles of operation of an SP Pack.

SP Packs are designed by the manufacturer to coalesce oil droplets to a defined drop size distribution, with a d_{max} of 1000 μm. They can be created by sizing a series of short runs of pipe with a diameter sized to create a Reynold's number of 50,000 contains 6–10 short radius 180-degree bends. Each straight run of pipe should be 30–50 pipe diameters long. Increasing the d_{max} from a typical value of 250 μm in a normal inlet to 1000 μm significantly reduces the size of the skimmer required. In addition, the need for retention time in the skimmer is not as important, since coalescence has occurred prior to the skimmer. As a result, retention times in the skimmer may be reduced to the 3- to 10-min range.

SP Packs may be effectively multistaged, as shown in Figure 3.31. As shown in Figure 3.32, a two-stage system may consist of an SP Pack, a skim vessel, a second SP Pack, and a second identical skim vessel. One SP Pack and skimmer combination constitutes one stage of coalescence and separation. The second SP Pack coalesces the small oil droplets in the first skimmer's outlet; then the second skimmer may remove the larger oil droplets.

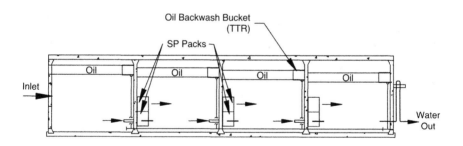

FIGURE 3.31. SP Pack installed in a horizontal flume.

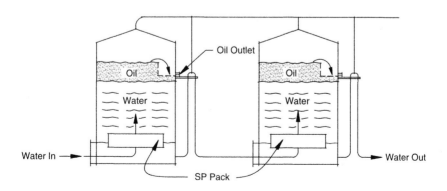

FIGURE 3.32. SP Packs installed in a series of staged tanks.

The addition of the SP Pack greatly improves the oil removal in the second skimmer because of coalescence. If the second SP Pack were not in the system, all the large oil droplets would be removed in the first skimmer and the second skimmer would remove little oil. Any number of stages in series may be used in the system.

SP Packs can also be used as retrofit components to improve the performance of existing water treating systems. Deck drainage may also be routed through an SP Pack prior to the drain sump or disposal pile.

SP Pack systems may be economical onshore where space is available for large skim tanks. Offshore SP Packs may be used for small water rates, roughly 5000 bwpd (33 m^3/h). If space is available offshore, larger flow-rate applications may prove economical.

As shown in Figure 3.33, the SP Pack is placed inside any gravity settling device (skimmers, clarifiers, plate coalescers, etc.), and by growing a larger drop size distribution, the gravity settler is more efficient at removing oil, as shown in Figure 3.34.

Performance Considerations
The efficiency in each stage is given by

$$E = \frac{C_i - C_o}{C_i},$$
(3.23)

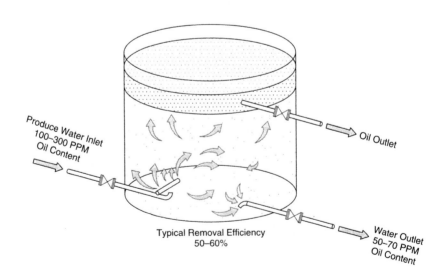

FIGURE 3.33. SP Pack installed in a clarifier skim tank.

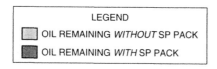

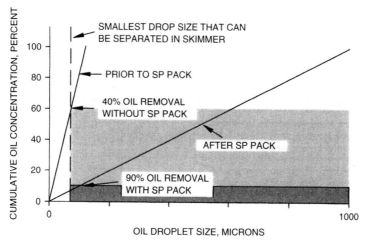

FIGURE 3.34. The SP Pack grows a larger droplet size distribution, thus allowing the skimmer to recover more oil.

where

C_i = inlet concentration,
C_o = outlet concentration.

Since the drop size distribution developed by the SP Pack can be conservatively estimated as a straight line,

$$E = 1 - \frac{d_m}{d_{max}}, \qquad (3.24)$$

where

d_m = drop size that can be treated in the stage,
d_{max} = maximum size drop created by the SP Pack
= 1000 μm for standard SP Packs.

The overall efficiency of a series staged installation is then given by

$$E_t = 1 - (1 - E)^n, \qquad (3.25)$$

where n is the number of stages.

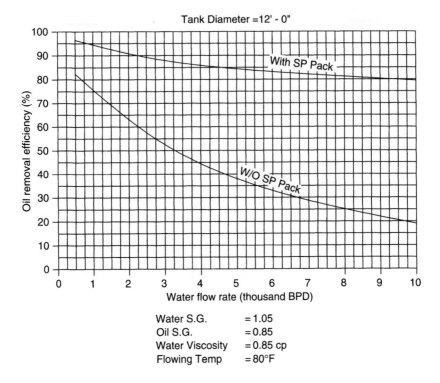

FIGURE 3.35. Improved oil removal efficiency of an SP Pack installed in a 12'-0' tank.

Figures 3.35 and 3.36 illustrate the increased oil removal efficiency of an SP Pack installed in various sized tanks.

3.6.6 Flotation Units

Flotation units are the only commonly used water treating equipment that does not rely on gravity separation of the oil droplets from the water phase. Flotation units employ a process in which fine gas bubbles are generated and dispersed in water, where they attach themselves to oil droplets and/or solid particulates. The gas bubbles then rise to the vapor–liquid interface as oily foam, which is then skimmed from the water interface, recovered, and then recycled for further processing. The effective specific gravity of the oil–gas bubble combination is significantly lower than that of a stand-alone oil droplet. According to Stokes' law, the resulting rising velocity of the oil–gas bubble combination is greater than that of a stand-alone oil droplet acting to accelerate the oil–water separation process. Flotation aids

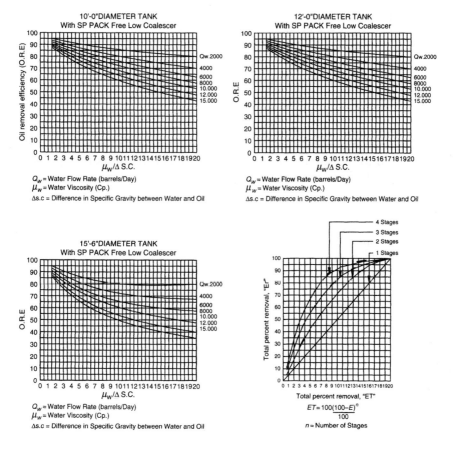

FIGURE 3.36. Oil removal efficiencies of various size tanks.

such as coagulants, polyelectrolytes, or demulsifiers are added to improve performance.

Two distinct types of flotation units have been used; they are distinguished by the method employed in producing the small gas bubbles needed to contact the water. These are dissolved gas units and dispersed gas units.

Dissolved Gas Units

Dissolved gas designs take a portion of the treated water effluent and saturate the water with natural gas in a high-pressure "contactor" vessel. The higher the pressure, the more gas that can be dissolved in the water. Gas bubbles are formed by flashing dissolved gas into the produced water. As a result, the bubbles are much smaller (10–100 µm) than for induced gas flotation (100–1000 µm). On the other hand, the gas volumes are limited by the solubility of the gas in water and are much lower than for dispersed gas flotation.

Most units are designed for a 20- to 40-psig (140- to 280-kPa) contact pressure. Normally, 20–50% of the treated water is recirculated for contact with the gas. The gas saturated water is then injected into the flotation tank as shown in Figure 3.37. The dissolved gas breaks out of the oily water solution when the water pressure is flashed (reduced) to the low operating pressure of the gas flotation unit, in small-diameter bubbles that contact the oil droplets in the water and bring them to the surface in froth. This type of flotation unit typically has not worked well in the oil field.

Dissolved gas units have been used successfully in refinery operations where air can be used as the gas, where large areas are available for the equipment, and where the water to be treated is, for the most part, oxygenated fresh water. In treating produced water, for injection, it is desirable to use natural gas to exclude oxygen, to avoid creating an explosive mixture, and to minimize corrosion and bacteria growth. This requires the venting of the gas or installation of a vapor recovery unit. In addition, the high dissolved solids content of produced water has created scale problems in these units. Field experiences with dissolved natural gas units in production operations have not been as successful as experience with dispersed gas units.

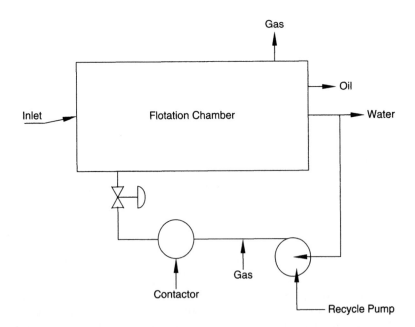

FIGURE 3.37. Schematic of a dissolved gas flotation process system.

Design parameters are recommended by the individual manufacturers but normally range from 0.2–0.5 scf/barrel (0.036–0.89 Sm^3/m^3) of water to be treated and flow rates of treated plus recycled water of between 2 and 4 gpm/ft^2 (4.8 and 9.8 m^3/m^3). Retention times of 10–40 min and depths of between 6 and 9 ft (1.8 and 2.7 m) are specified.

Dissolved gas units are common in chemical plant operations, but, for the following reasons, they are seldom used in producing operations:

- They are larger than dispersed gas units and they weigh more, so they have limited application offshore.
- Many production facilities do not have vapor recovery units and, thus, the gas is not recycled.
- Produced water has a greater tendency to cause scale in the bubble-forming device than the freshwater that is normally found in plants.

Dispersed Gas Units

In dispersed gas units, gas bubbles are dispersed in the total stream either by the use of a hydraulic inductor device or by a vortex set up by mechanical rotors. There are many different proprietary designs of dispersed gas units. All require a means to generate gas bubbles of favorable size and distribution into the flow stream, a two-phase mixing region that causes a collision to occur between the gas bubbles and the oil droplets, a flotation or separation region that allows the gas bubbles to rise to the surface, and a means to skim the oily froth from the surface. Figure 3.38 shows the regions in which the above four processes occur. These processes and the regions in which they occur are as follows:

- Gas circulation path (A) and fluid circulation path (B) (bubble generation),
- Two-phase mixing region (1) (attachment of oil droplets to the bubbles),
- Flotation (separation) region (2) (rise of the bubbles to the surface applying Stokes' law),
- Skimming region (3) (bubble collapse and oil skimming).

Gas bubble/oil droplet attachment can be enhanced with the use of polyelectrolyte chemicals. These flotation aid chemicals can also be used to cause bubble/solid attachments, and thus flotation units can be used to remove solids as well. These chemicals are typically added to the water to yield a chemical concentration level between

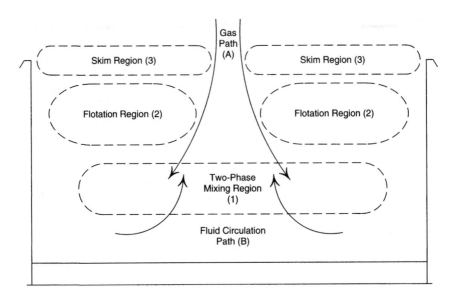

FIGURE 3.38. Dispersed gas flotation cell mechanics.

1 and 10 ppm in the feed water. These chemicals are classified as sur-factants, which tend to migrate to the bubble surface where they enhance the intermolecular forces between the bubbles and the oil droplets. In addition, if the oil has emulsifying tendencies, de-emulsi-fiers may also have to be added, in the 20- to 50-ppm range. Chemical treatment programs are highly location specific, and an effective treatment for one oil–water system may be ineffective for another. An on-site bench-scale flotation chemical screening test using a pilot flotation device should be carried out for each application. Equipment manufacturers and chemical suppliers are generally equipped to per-form this screening.

For the oil droplets to become attached to the bubbles, the bub-bles and droplets must come into intimate contact. This contact is promoted by a highly turbulent region, generally located near the bub-ble generators. Studies have shown that attachment is enhanced by small gas bubbles, large oil droplets, and high bubble concentration.

To operate efficiently, the unit must generate a large number of small gas bubbles. Tests indicate that bubble size decreases with increasing salinity. At salinities above 3%, bubble size appears to remain constant, but oil recovery often continues to improve. Most oil-field waters contain sufficient dissolved solids to create favorable flotation bubble sizes. The low water salinity associated with gas con-densates may make the application of gas flotation to gas condensate fields more difficult than for oil fields. Some steam-flood produced waters contain 2000–5000 ppm of dissolved salts and would tend to generate large, less effective bubbles.

Figure 3.39 shows the effect of gas bubble size on the oil droplet capture rate. The smaller the bubble size, the greater the chance of capture will be. Typical mean bubble sizes range between 50 and 60 μm.

Oil removal is dependent to some extent on oil droplet size. Flotation has very little effect on oil droplets that are smaller in diameter than 2–5 μm. Thus, it is important to avoid subjecting the influent to large shear forces (e.g., level control valves) immediately upstream of the unit. It is best to separate control devices from the unit by long lengths of piping (at least 300 diameters) to allow pipe coalescence to increase droplet diameter before flotation is attempted. Above 10–20 μm, the size of the oil droplet does not appear to affect oil recovery efficiency, and thus elaborate inlet coalescing devices are not needed.

High gas bubble concentration (fraction of the gas–water mixture, i.e., vapor) increases the oil recovery. High gas bubble concentrations cannot be obtained in the presence of small oil droplets. Thus, choking or pumping the produced water in a fashion that would create high shear forces and the formation of small oil droplets should be avoided. Field tests demonstrate that oil removal improves as the

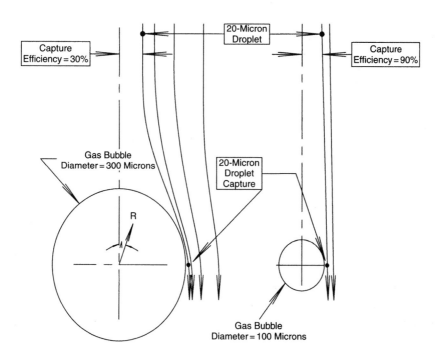

FIGURE 3.39. Effect of gas bubble size on oil droplet rate.

TABLE 3.3
Effects of increased gas concentration on oil recovery

Water Locations in Machine	*Cumulative Gas–Water Ratio, ft³/bbl*	*PPM Oil in Treated Water*
Inlet water	0	225
Cell no. 1 effluent	9	96
Cell no. 2 effluent	18	50
Cell no. 3 effluent	27	20
Cell no. 4 effluent	34	14
Discharge cell effluent	35	14

cumulative gas–water ratio increases. Table 3.3 illustrates the effects of increased gas concentration on oil recovery.

The rise and separation of oily bubbles from water require a relatively quiescent zone so that bubbles are not remixed into the bulk fluid. The rising velocity of the bubbles must exceed both turbulent velocities and any net downward bulk velocity.

The oily bubbles rise to the surface as an oily foam, which is then skimmed from the surface of the bulk water phase. This skimming process acts to collapse the foam, which further concentrates the oily phase. Skimming is usually achieved by a combination of weirs and skim paddles that move the oily foam to the edge of the cell and over the weir. The weir height relative to the position and speed of the skim paddles must be adjusted to prevent both excessive foam build-up on the bulk water surface and excessive water carryover into the oil skim bucket located below the weir.

Hydraulic Induced Units
Hydraulic induced flotation units induce gas bubbles by gas aspiration into the low-pressure zone of a venture tube. Figure 3.40 shows a schematic cross section of a hydraulic induced flotation unit. Clean water from the effluent is pumped to a recirculation header (E) that feeds a series of Venturi eductors (B). Water flowing through the eductors sucks gas from the vapor space (A) that is released at the nozzle (G) as a jet of small bubbles. The bubbles rise, causing flotation in the chamber (C), forming a froth (D) that is skimmed with a mechanical device at (F).

Hydraulic induced units are available with one, three, or four cells. Figure 3.41 shows the flow path through a three-cell unit. These devices use less power and less gas than mechanical rotor units. Gas–water ratios less than 10 ft²/bbl at the design flow rate are used. The

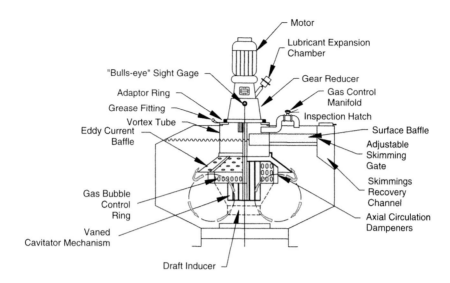

FIGURE 3.40. Schematic of a hydraulic induced gas flotation unit.

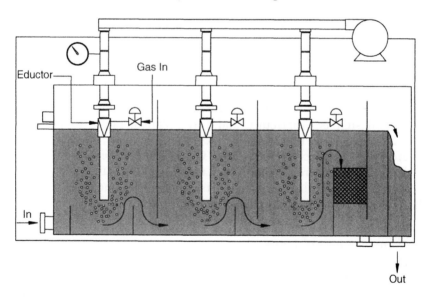

FIGURE 3.41. Schematic showing the flow path through a hydraulic induced flotation unit.

volume of gas dispersed in the water is not adjustable, so throughputs less than design result in higher gas–water ratios.

Hydraulic induced units are less complex than the mechanical induced units. In contrast to the mechanical units, the hydraulic units are capable of operating the flotation process above atmospheric

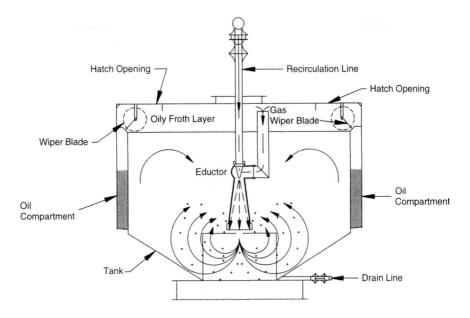

FIGURE 3.42. Cross section of a hydraulic inductor.

pressure. The motor-driven paddles and bubble distributors are replaced by a Venturi-type eductor. The required water recycle rate to drive the eductor varies with both the design capacity of the unit and between different manufacturers, but is generally around 50%. Eductor design is proprietary and varies considerably, in both hydraulic design and mechanical placement between manufacturers. Figure 3.42 is a sketch of an eductor. Control of bubble size and distribution is much more difficult than for mechanical units. Stage efficiencies for hydraulic induced units have a tendency to be lower than those of mechanical units.

Mechanical Induced Units

Mechanical induced flotation units induce gas bubbles into the system by entrainment of gas in a vortex generated by a stirred paddle. Figure 3.43 shows a cross section of a dispersed gas flotation cell that utilizes a mechanical rotor. The rotor creates a vortex and vacuum within the vortex tube. Shrouds assure that the gas in the vortex mixes with and is entrained in the water. The rotor and draft inducer causes the water to flow as indicated by the arrows in this plane while also creating a swirling motion. A baffle at the top directs the froth to a skimming tray as a result of this swirling motion.

Most dispersed gas units contain three or four cells. Figure 3.44 illustrates a four-cell unit. Bulk water moves in series from one cell

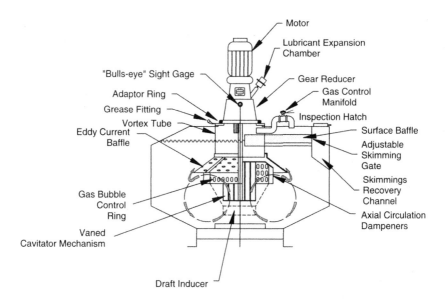

FIGURE 3.43. Cross section of a mechanical induced dispersed gas flotation unit.

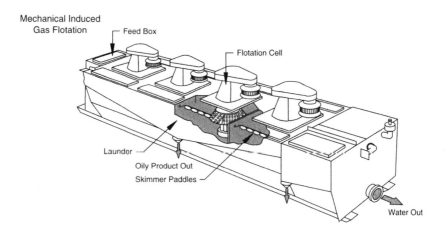

FIGURE 3.44. Cross section of a four-cell mechanical induced flotation unit.

to the other by underflow baffles. Each cell contains a motor-driven paddle and associated bubble generation and distribution hardware. Field tests have indicated that the high intensity of mixing in each cell creates the effect of plug flow of the bulk water from one cell to the next. That is, there is virtually no short-circuiting or breakthrough of a part of the inlet flow to the outlet weir box.

The mechanical complexity makes mechanical induced flotation units the most maintenance-intensive of all gas flotation configurations. As a result of the need for motor shaft seals on penetrations to the cell, mechanical induced flotation units have traditionally operated very near atmospheric pressure.

Each of the above processes assumes the inlet water to the flotation unit is already at atmospheric pressure. When the upstream primary separator operates at elevated pressures, substantial gas saturation of the produced water may already exist. In these cases, flashing to atmospheric pressure may be sufficient to generate bubbles without added gas saturation.

Other Configurations

The combination of dissolved gas flotation and CPIs has been attempted recently, with injection of a recycled portion of the effluent from the CPI into the influent stream. Little field data are available on this design; therefore, it is not recommended at this time because, when the dissolved air breaks out of solution, turbulence that can adversely affect the action of the CPI is created.

There are many types of configurations with complex flow patterns and number of cells ranging from one to five. Some designs have multiple eductors per cell. Some have recirculation rates through the eductors that may be several multiples of the bulk water throughput rate. The concept described above should give the engineer some guidance to add in understanding the pros and cons of each manufacture's proprietary designs.

One new design when shows some promise uses a "Sparger." A Sparger introduces gas from an external high pressure source similar to that of an aerator in an aquarium. Porous media nozzles are used to form very small bubbles, the size of which is controlled somewhat by the pore size in the media. For the most effective attachment of oil droplets to these sparged bubbles, the bubble size should be approximately the same as the smallest oil droplets to be removed. Due to the small bubbles that are used in sparging, long fluid residence times on the order of 10 minutes are required. Sparging generates some mechanical complexity due to the needs for a separate pressurized gas supply and for numerous porous media nozzles that may be prone to plugging. On the other hand, using multiple spargers could generate smaller bubbles, greater flow rates and better gas mixing with the produced water than other designs. The detriment is that the porous media could plug with time leading to high maintenance costs and poor availability.

Sizing Dispersed Gas Units

It can be shown mathematically that an efficient design must have a high gas induction rate, a small-diameter induced gas bubble, and a relatively large mixing zone. The design of the nozzle or rotor, and of the internal baffles, is thus critical to the unit's efficiency. The nozzles, rotors, and baffles for these units are patented designs.

As measured in actual field tests, these units operate on a constant percent removal basis. Within normal ranges their oil removal efficiency is independent of inlet concentration or oil droplet diameter.

Field tests indicate that a properly designed unit with a suitable chemical treatment program should have oil removal efficiency between 40% and 55% per active cell and an overall efficiency of about 90%. An excellently designed system might exhibit an efficiency as high as 95%, while a poorly designed, poorly operated unit, or difficult oil–water chemistry could easily degrade performance to as low as 80%. Equation (3.25) verifies the above efficiencies. For example, Equation (3.25) shows that a three-cell unit can be expected to have an overall efficiency of 87% while a four-cell unit can be expected to have an overall efficiency of 94%. The unit's actual efficiency will depend on many factors that cannot be controlled or predicted in laboratory or field tests.

Each cell is designed for approximately 1-min retention time to allow the gas bubbles to break free of the liquid and form the froth at the surface. Each manufacturer gives the dimensions of its standard units and the maximum flow rate based on this criteria.

Graphs of the dispersed oil concentrations in the effluent water versus dispersed oil concentrations in the inlet feed stream are shown in Figure 3.45 for representative efficiencies achievable in a typical four-cell dispersed gas flotation unit. For inlet concentrations less than about 200 mg/l, the oil removal efficiency declines slightly. At low oil inlet concentrations, it becomes more difficult for the flotation unit to achieve intimate contact and interaction between the gas bubbles and dispersed oil droplets. As a result, Figure 3.45 may understate the effluent concentrations for influent oil concentrations less than 200 mg/l.

Depending on the oil concentration in the influent and the quality requirements in the effluent, flotation may or may not serve as a standalone process in produced water treating. Water qualities coming from a primary production separator tend to be in the 500- to 2000-ppm range. As can be seen from Figure 3.45, a well-designed gas flotation unit would be limited to an effluent quality in the 30- to 80-ppm range when used as the sole water treating unit downstream of the primary separator. Since the separation efficiency is reasonably independent

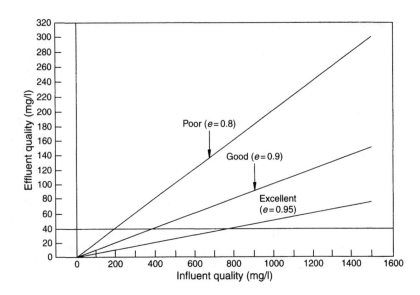

FIGURE 3.45. Effluent quality versus influent quality.

of the influent oil concentration, upsets in the primary separator operation could make a significant difference in the gas flotation effluent quality. In order to meet the effluent quality established by the authorities having jurisdiction, normally in the 30- to 50-ppm range, it is usually necessary to combine a gas flotation unit with some unit between the flotation unit and the primary separator, such as a CPI.

Gas flotation units require a customized chemical treatment program to achieve adequate results. If produced water originates from several sources in variable quantities, the development of a chemical treatment program may be difficult.

Skimmed oily water volumes are typically 2–5% of the machine's rated capacity and can be as high as 10% when there is a surge of water flow into the unit. Since skimmed fluid volume is a function of weir length exposure over time, operation of the unit at less than design capacity increases the water residence time but does not decrease the skimmed fluid volumes.

Gas flotation units normally include multiple cells in series. If the mechanical or hydraulic aeration unit in any cell fails, the water merely flows through that cell with little or no oil separation. As a result, mechanical failure in a single cell causes a degradation in performance. For example, a four-cell unit with a mechanical failure in one cell becomes a three-cell unit capable of separating 87.5% of the dispersed oil in the feed stream (i.e., $1 - (1 - 0.5)^3 = 0.875$).

Gas flotation equipment is typically purchased as a prefabricated unit selected from a vendor's list of standard size units, rather than being custom specified and designed for each specific application. As a result the bulk of the design is performed by the vendor, and relatively little design opportunity exists for the user.

Performance Considerations

Several factors that should be taken into account to maintain performance include:

- The cells must be properly leveled on initial installation, and this level condition must be maintained. Since the skimming depends on proper operation of a weir, small out-of-level conditions will prevent proper skimming of oil. Movement of the flotation cells can also set up liquid surges that can prevent proper skimming.

- Liquid levels must be carefully controlled to permit proper weir operation. Level control system parameters must be carefully set to prevent liquid level oscillations. Throttling valves are preferred over snap acting valves on both the water inlet and outlet. The flow disturbances caused by the rapid opening and closing of the snap acting valves may generate level disturbances. Gravity flow of the inlet feed stream to the gas flotation unit is preferred over pumping. The high shearing action created in a pump will break up the larger oil droplets into smaller droplets, making separation more difficult.

- Many induced flotation units, particularly mechanical flotation units, operate at pressures within a few ounces of atmospheric pressure. The walls are thin and have numerous penetrations for motor shafts and observation hatches. As a result of the simplicity of design, air can easily enter the units around the paddle or if observation hatches are left open. Oxygen in the water treating system increases the corrosion rate in the unit as well as all downstream carbon steel equipment and can cause the formation of a reddish precipitate resulting from oxidation of dissolved iron in the treated water. To avoid corrosion and the precipitate, care should be taken to avoid oxygen ingress. Hatches should be left closed as much as possible, and the integrity of shaft seals should be maintained.

- Proper chemical treatment is essential to the operation of gas flotation. Care must be taken to ensure that the chemical injection facilities are operating as expected and that proper dosage is administered and mixed, both to promote sufficient separation and to prevent excessive chemical use. The customized chemical treatments involving polyelectrolytes, de-emulsifiers, scale inhibitors, and corrosion inhibitors may result in chemical incompatibilities, either between chemicals or between chemical treatments and flotation cell materials. These incompatibilities may be compounded by propagation through the treatment facility of any chemicals added upstream of the water treating system. Units should be monitored for any unexpected sludge or precipitates or for unexpectedly high corrosion or elastomer deterioration rates.

- Stripping of acid gases (H_2S, CO_2) in an induced gas flotation unit can cause a pH increase, which may result in scaling. If this is the case, minimizing the gas flow can help reduce this problem.

Field tests indicate the following performance findings:

- Induced gas flotation units remove 100% of oil droplets 10–20 μm and above and have some effect on oil droplets in the 2- to 5-μm range.
- Oil removal efficiency depends on chosing the correct chemical and dosage. (refer to Figure 3.46).
- Mechanical units tend to be more efficient than hydraulic units with influent concentrations from 50 to 150 mg/l, while hydraulic units are more efficient above 500 mg/l. Both units are equally efficient between 150 and 500 mg/l.
- The performances of all induced gas flotation units are relatively insensitive to flow-rate variations between 70% and 125% of the design flow rate.
- Mechanically induced units appear to tolerate greater throughout rate fluctuations than hydraulic induced units.
- The separation efficiency of all units depends on the influent concentration.
- Changing the water temperature from ambient to 140 °F (60 °C) results in a slight improvement in oil recovery at normal pH values.

Gas flotation units should be used when:

- The inlet oil concentrations are not too high (250–500 mg/l).
- The effluent discharge requirements are not too severe (25–50 mg/l).
- Chemical companies are available to formulate an appropriate chemical treatment program.
- Power costs are low or moderate.
- Oil/water density differences are low, such as heavy oils.

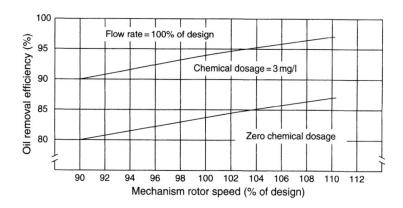

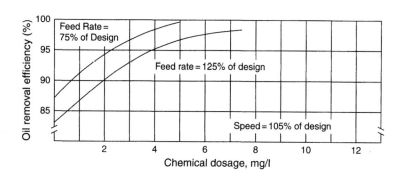

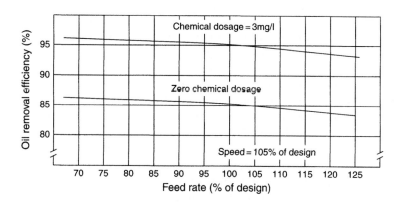

FIGURE 3.46. Oil removal efficiencies of mechanical induced flotation units.

Gas flotation units should not be used when:

- Equipment size and weight are prime considerations.
- The unit is subject to accelerations and tilting such as floating production facilities.
- The water stream to be treated is comprised of multiple water sources having significantly varying water chemistry and dispersed oil characteristics.
- Service support from water treating chemical vendors is limited.
- Very low effluent oil concentrations are required.
- Power costs are high.

3.6.7 Hydrocyclones

General Considerations

Since the early 1980s, hydrocyclones have been used in produced water treatment to de-oil the water prior to discharge. Hydrocyclones used to de-oil the water are referred to as "liquid–liquid de-oiling" hydrocyclones. Liquid–liquid hydrocyclones are further classified as static or dynamic hydrocyclones.

Operating Principles

Hydrocyclones, sometimes called "enhanced gravity separators," use centrifugal force to remove oil droplets from oily water. As shown in Figure 3.47, a de-oiling static hydrocyclone typically consists of liner(s) contained within a pressure retaining outer vessel or shell. The liner consists of the following four sections: a cylindrical swirl chamber, a concentric reducing section, a fine tapered section, and a cylindrical tail section. Figure 3.48 shows a typical multiliner vessel.

Oily water enters the cylindrical swirl chamber through a tangential inlet nozzle (Figure 3.49), creating a high-velocity vortex with a reverse-flowing central core. The fluid accelerates as it flows through the concentric reducing section and the fine tapered section. The fluid then continues at a constant rate through the cylindrical tail section. Larger oil droplets are separated from the fluid in the fine tapered section, while smaller droplets are removed in the tail section. Centripetal forces cause the lighter-density droplets to move toward the low-pressure central core, where axial reverse flow occurs. The oil is removed through a small-diameter reject port located in the head of the hydrocyclone. Clean water is removed through the downstream outlet.

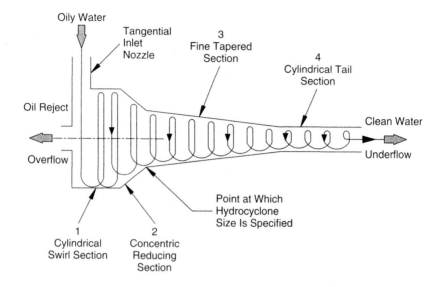

FIGURE 3.47. Liquid–liquid static hydrocyclone separation liner.

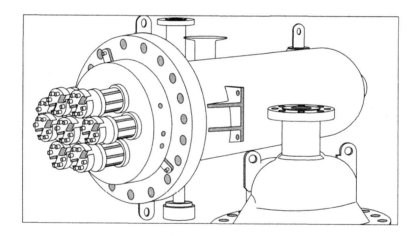

FIGURE 3.48. Multiliner hydrocyclone vessel.

The separation mechanism inside a hydrocyclone is governed by Stokes' law. However, in a hydrocyclone, the gravitational force is orders of magnitude (between 1000–2000 gs) higher than that available in conventional gravity-based separation equipment. A high-velocity vortex with a reverse-flowing central core is set up by entry through a specially designed tangential inlet(s) (Figure 3.49). The fluid is accelerated (thereby offsetting the frictional losses) through the concentric reducing and fine tapered sections of the cyclone—where the bulk of

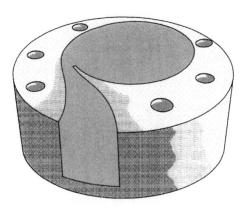

FIGURE 3.49. Tangental inlet nozzle.

separation occurs—into the cylindrical tail section where smaller, slower-moving droplets are recovered. The size of a hydrocyclone (for example, 35 or 60 mm) refers to the diameter at transition between the concentric reducing and fine tapered section of the cyclone (see Figure 3.47).

The lighter oil droplets migrate toward the inner central core where an axial reversal of flow occurs, resulting in the removal of the lower-density oil-enriched phase through a small-diameter orifice (the reject port or the vortex finder) located in the center of the inlet head. This stream is also known as the reject stream or the overflow. The oil-depleted water stream exits from the downstream end (also known as underflow).

A hydrocyclone can be oriented either horizontally or vertically although horizontal orientation is more common. The horizontal orientation requires more plan area (deck space) but is more convenient for maintenance (about a 42-in. clearance is required to remove the liners from the vessel). The energy required to achieve separation is provided by the differential pressure across the cyclone. A minimum of 100 psi is generally needed. Higher pressures are preferable when available. The reject stream is on the order of 1–3 vol.% of the inlet. Only about 10% (by volume) of the reject stream is oil, the rest being water. The reject stream may be directed back to the separator through a low-shear progressive cavity pump. It should be noted, however, that in certain field applications oil-field chemicals have caused swelling of the rubber stator of these pumps, leading to poor performance. In such situations, a low-speed single-stage centrifugal pump with an open impeller may be suitable.

Many hydrocyclone installations typically include a degassing vessel downstream of the clean product water outlet. The vessel provides a short residence time serving essentially as a single gas flotation unit. The vessel also provides oil slug-catching volume in case of major upsets and additional residence time for emulsion-breaking chemicals.

Static Hydrocyclones

Static hydrocyclones require a minimum pressure of 100 psi to produce the required velocities. Manufacturers make designs that operate at lower pressures, but these models have not always been as efficient as those that operate at higher inlet pressures. If a minimum separator pressure of 100 psi is not available, a low-shear pump should be used (e.g., a progressive cavity pump) or sufficient pipe should be used between the pump and the hydrocyclone to allow pipe coalescence of the oil droplets. As is the case with flotation units, hydrocyclones do not appear to work well with oil droplets less than 10–20 μm in diameter.

Performance is chiefly influenced by the reject ratio and the pressure drop ratio (PDR). The reject ratio refers to the ratio of the reject fluid rate to the total inlet fluid rate. Typically, the optimum ratio is between 1% and 3%. This ratio is also proportional to the PDR. Operation below the optimum reject ratio will result in low oil removal efficiencies. Operation above the optimum reject ratio does not impair oil removal efficiency, but it increases the amount of liquid that must be reticulated through the facility. The PDR refers to the ratio of the pressure difference between the inlets and reject outlets and the difference from the inlet to the water outlet. A PDR of between 1.4 and 2.0 is usually desired. Performance is also affected by inlet oil droplet size, concentration of inlet oil, differential specific gravity, and inlet temperature. Temperatures greater than 80 °F result in better operation.

Although the performance of hydrocyclones varies from facility to facility (as with flotation units), an assumption of 90% oil removal is a reasonable number for design. Often the unit will perform better than this, but for design it would be unwise to assume this will happen. Performance cannot be predicted more accurately from laboratory or field testing because it is dependent on the actual shearing and coalescing that occur under field flow conditions and on impurities in the water, such as residual treating and corrosion chemicals and sand, scale, and corrosion products, which vary with time.

Hydrocyclones are excellent coalescing devices, and they actually function best as a primary treating device followed by a downstream skim vessel that can separate the 500- to 1000-μm droplets that leave with the water effluent. A simplified P&ID for a hydrocyclone is shown in Figure 3.50.

Advantages of static hydrocyclones include that (1) they have no moving parts (thus, minimum maintenance and operator attention are required), (2) their compact design reduces weight and space

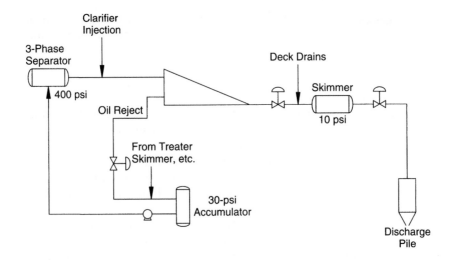

FIGURE 3.50. Simplified P&ID showing a hydrocyclone used as preliminary treating device.

requirements when compared to those of a flotation unit, (3) they are insensitive to motion (thus, they are suitable for floating facilities), (4) their modular design allows easy addition of capacity, and (5) they offer lower operating costs when compared to flotation units, if inlet pressure is available.

Disadvantages include the need to install a pump if oil is available only at low pressure and the tendency of the reject port to plug with sand or scale. Sand in the produced water will cause erosion of the cones and increase operating costs.

Performance of hydrocyclones is affected by the following parameters:

- *Oil droplet size (at fixed concentrations).* Efficiency generally decreases as the oil droplet size is reduced. This is consistent with Stokes' law, where a smaller droplet will move less rapidly toward the hydrocyclone core. Droplets below a certain size (about 30 µm) are not captured by the hydrocyclone and, therefore, as the median feed oil droplet size decreases, more of the smaller droplets escape and the efficiency drops. Restrictions (valves, fittings, etc.) and pumps causing droplet shearing in the incoming flow should be avoided.
- *Differential specific gravity.* At a constant temperature, the hydrocyclone oil removal efficiency increases as the salinity increases and/or the crude specific gravity decreases. As the specific gravity difference between water and oil increases, a greater driving force for oil removal in the hydrocyclone occurs.

- *Inlet temperature.* The temperature of the produced water inlet stream determines the viscosity of the oil and water phases and the density difference between the two phases. As temperature increases, the viscosity of water decreases slightly, while the density difference increases more substantially. This is because oil density decreases at a faster rate than water density. Temperatures greater than 80 °F result in better operation.
- *Inlet flow rate.* The centrifugal force induced in the hydrocyclone is a function of the flow rate. At low flow rates, insufficient inlet velocity exists to establish a vortex and separation efficiency is low. Once the vortex is established, the efficiency increases slowly as a function of the flow rate to a point where the pressure at the core approaches atmospheric. Any further increase in the flow rate hinders oil flow from the reject outlet and causes efficiency to decline. In addition, a high flow rate can cause shearing of the droplets. This maximum flow rate is the "capacity" of the chamber. Flow rate is controlled by back pressure on the underflow outlet. The ratio of maximum to minimum flow rate, as determined by the lowest separation efficiency acceptable and the available pressure drop, is the "turndown ratio" for a given application.

Dynamic Hydrocyclones

The major difference between static and dynamic hydrocyclones is that in the dynamic hydrocyclone an external motor is used to rotate the outer shell of the hydrocyclone, whereas in a static hydrocyclone the outer shell is stationary and feed pressure supplies the energy to accomplish separation of oil from water (no external motor is required).

As shown in Figure 3.51, a dynamic hydrocyclone consists of a rotating cylinder, axial inlet and outlet, reject nozzle, and external motor. The rotation of the cylinder creates a "free vortex." The tangential speed is inversely proportional to the distance to the centerline of the cyclone. Since there is no complex geometry that requires a high pressure drop, dynamic units can operate at lower inlet pressures (approximately 50 psig) than static units. In addition, the effect of the reject ratio is not as important in dynamic units as it is in static units.

Dynamic hydrocyclone performance is affected by the following parameters:

- *Reject flow.* The reject flow must remain constant across the inlet flow range.
- *Rotational speed.* High rotational speeds (between 1000 and 3000 rpm) generate higher centrifugal forces, which in turn yield better oil removal efficiencies at a given flow rate.

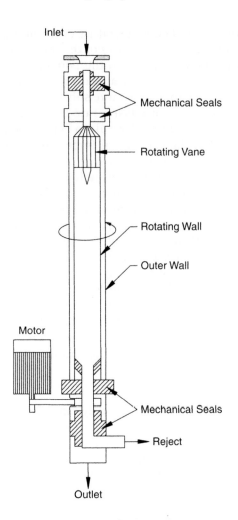

FIGURE 3.51. Liquid–liquid dynamic hydrocyclone separation.

The oil removal efficiency, which is a function of the drop size distribution, is between 50% and 75%. Dynamic units are more effective at removing small oil droplets (15 μm) from the water.

Dynamic hydrocyclones have found few applications because of the poor cost–benefit ratio.

Selection Criteria and Application Guidelines
Hydrocyclones can be applied when:

1. Median oil particle size is in excess of 30 μm.
2. Produced water feed pressure is at least 100 psig.

3. Platform deck space is a critical consideration. Hydrocyclones have low size and weight requirements compared to other water treating equipment for the same capacity.
4. Platform motion is significant, such as tension leg platforms or floating production facilities, since the hydrocyclone is insensitive to motion. Other devices, such as flotation cells, are adversely impacted since platform motion makes accurate level control (using weirs or other control devices) difficult.
5. An appreciable quantity of solids is not present.
6. An appreciable amount of free gas is not present.
7. The flow rate and feed water oil concentration are fairly constant.
8. Low equipment maintenance is desired. Since a hydrocyclone has no moving parts, its maintenance requirements are fairly low.
9. Power constraints exist. Hydrocyclones do not require any outside energy supply, except for a low HP (about 5 HP) reject–recycle pump.

They are not applicable when:

1. A tight emulsion exists, with a median oil droplet size less than 30 μm (manufacturer's claim that newer high-efficiency liners are capable of removing 20 μm).
2. The feed water pressure is less than 100 psig. A pump would be required to develop adequate pressure to use a hydrocyclone. However, pumps can cause the oil droplets to shear, making it more difficult for separation by a hydrocyclone.
3. The difference in specific gravity between the oil and water is relatively low, that is, heavy crude is being produced.
4. Considerable sand is entrained in the produced water. The sand could potentially plug the reject orifice and also cause erosion of the liner.

Sizing and Design
The performance of hydrocyclones is measured in terms of oil removal efficiency (E).

$$\text{Product oil removal efficiency:} \ E = \frac{[(C_i - C_o)(100)]}{C_i},$$

where

C_i = dispersed oil concentration in feed water,
C_o = dispersed oil concentration in effluent water.

Figure 3.52 shows generalized removal efficiency curves of a hydrocyclone. For a typical case (30 °API oil and 1.05 SG water), the differential specific gravity is 0.17 and the removal efficiency would be 92% of 40-μm, 85% of 30-μm, and 68% of 20-μm oil droplets.

Figure 3.53 shows a typical control scheme for a hydrocyclone.

3.6.8 Disposal Piles

Disposal piles are large-diameter (24- to 48-in.), open-ended pipes attached to the platform and extending below the surface of the water. Their main uses are to (1) concentrate all platform discharges into one location, (2) provide a conduit protected from wave action so that discharges can be placed deep enough to prevent sheens from occurring during upset conditions, and (3) provide an alarm or shutdown point in the event of a failure that causes oil to flow overboard.

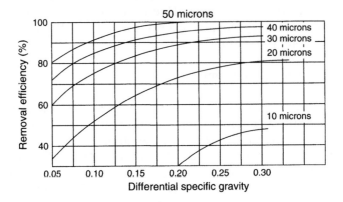

FIGURE 3.52. Generalized performance curves for hydrocyclone.

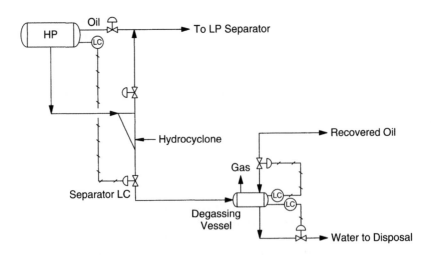

FIGURE 3.53. Typical control scheme for a hydrocyclone.

Most authorities having jurisdiction require all produced water to be treated (skimmer tank, coalesced, or flotation) prior to disposal in a disposal pile. In some locations, disposal piles are permitted to collect treated produced water, treated sand, liquids from drip pans and deck drains, and as a final trap for hydrocarbon liquids in the event of equipment upset.

Disposal piles are particularly useful for deck drainage disposal. This flow, which originates either from rainwater or wash-down water, typically contains oil droplets dispersed in an oxygen-laden freshwater or saltwater phase. The oxygen in the water makes it highly corrosive, and commingling with produced water may lead to scale deposition and plugging in skimmer tanks, plate coalescers, or flotation units. The flow is highly irregular and would thus cause upsets throughout these devices. Finally, this flow must gravitate to a low point for collection and either is pumped up to a higher level for treatment or treated at that low point. Disposal piles are excellent for this purpose. They can be protected from corrosion, they are by design located low enough on the platform to eliminate the need for pumping the water, they are not severely affected by large instantaneous flow-rate changes (effluent quality may be affected to some extent, but the operation of the pile can continue), they contain no small passages subject to plugging by scale build-up, and they minimize commingling with the process since they are the last piece of treating equipment before disposal.

Disposal Pile Sizing

The produced water being disposed of has been treated in vessels having the capability of treating smaller droplets than those that can be predicted to settle out in the relatively slender disposal pile. Small amounts of separation will occur in the disposal pile due to coalescence in the inlet piping and in the pipe itself. However, no significant treating of produced water can be expected.

Most authorities having jurisdiction require that deck drainage be disposed of with no free oil. If the deck drainage is merely contaminated rainwater, the disposal pile diameter can be estimated from the following equation, assuming the need to separate 150-μm droplets:

Field units

$$d^2 = \frac{0.3(Q_w + 0.356A_DR_w + Q_{WD})}{(\Delta SG)},$$ (3.26a)

SI units

$$d^2 = \frac{28,289(Q_w + 0.001A_DR_w + Q_{WD})}{(\Delta SG)},$$ (3.26b)

where

> d = pile internal diameter, in. (mm),
> Q_w = produced water rate (if in disposal pile), bwpd (m^3/h),
> A_D = plan area of the deck, ft^2 (m^2),
> R_w = rainfall rate, in./h (mm/h)
>   = 2 in./h for Gulf of Mexico (50 mm/h for GOM),
> ΔSG = difference in specific gravity between oil droplets and rainwater,
> Q_{WD} = wash-down rate, BPD (m^3/h)
>   = 1500 N (9.92 N),
> N = number of 50-gpm (189.25-l/min) wash-down hoses,

In Equations (3.26a) and (3.26b) either the wash-down rate or the rainfall rate should be used as it is highly unlikely that both would occur at the same time. The produced water rate is only used if produced water is routed to the pile for disposal.

The disposal pile length should be as long as the water depth permits in shallow water to provide for maximum oil containment in the event of a malfunction and to minimize the potential appearance of any sheen. In deep water the length is set to assure that an alarm and then a shutdown signal can be measured before the pile fills with oil.

These signals must be high enough so as not to register tide changes. The length of pile submergence below the normal water level required to assure that a high level will be sensed before the oil comes within 10 ft of the bottom is given by

Field units

$$L = \frac{(H_T + H_S + H_A + H_{SD})\text{SG}_o}{(\Delta\text{SG})} + 10, \qquad (3.27a)$$

SI units

$$L = \frac{(H_T + H_S + H_A + H_{SD})\text{SG}_o}{(\Delta\text{SG})} + 0.6, \qquad (3.27b)$$

where

> L = depth of pile below mean water level (MWL; submerged length), ft (m),
> H_t = normal tide range, ft (m),
> H_s = design annual storm surge, ft (m),
> H_A = alarm level (usually 2 ft), ft (m),
> H_{SD} = shutdown level (usually 2 ft or 0.6 m), ft (m),
> $(\text{SG})_o$ = specific gravity of the oil relative to water.

It is possible in shallow water to measure the oil–water interface for alarm or shutdown with a bubble arrangement and a shorter pile.

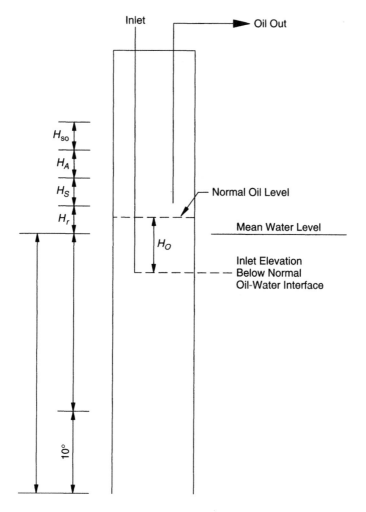

FIGURE 3.54. Disposal pile length.

However, this is not recommended where water depth permits a longer pile. To minimize wave action effects a minimum pile length of about 50 ft is required. Figure 3.54 is a schematic showing disposal pile length.

Skim Piles

The skim pile is a type of disposal pile. As shown in Figure 3.55, flow through the multiple series of baffle plates creates zones of no flow that reduce the distance a given oil droplet must rise to be separated from the main flow. Once in this zone, there is plenty of time for

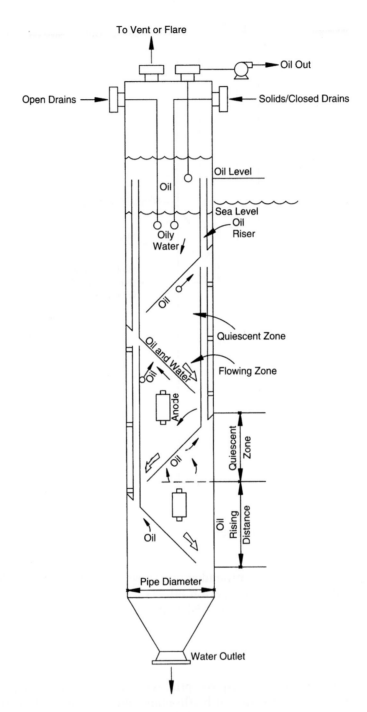

FIGURE 3.55. Cross section showing flow pattern of a skim pile.

coalescence and gravity separation. The larger droplets then migrate up the underside of the baffle to an oil collection system.

Besides being more efficient than standard disposal piles, from an oil separation standpoint, skim piles have the added benefit of providing for some degree of sand cleaning. Most authorities having jurisdiction state that produced sand must be disposed of without "free oil." It is doubtful that sand from a vessel drain meets this criterion when disposed of in a standard disposal pile.

Sand traversing the length of a skim pile will abrade on the baffles and be water-washed. This can be said to remove the free oil that is then captured in a quiescent zone.

Skim Pile Sizing

The determination of skim pile length is the same as that for any other disposal pile. Because of the complex flow regime, a suitable equation has yet to be developed to size skim piles for deck drainage. However, field experience has indicated that acceptable effluent is obtained with 20 min' retention time in the baffled section of the pile. Using this and assuming that 25% of the volume is taken up by the coalescing zones, we have the following:

Field units

$$d^2L' = 19.1(Q_w + 0.356A_D R_w + Q_{WD}), \tag{3.28a}$$

SI units

$$d^2L' = 565,811(Q_w + 0.001A_D R_w + Q_{WD}), \tag{3.28b}$$

where L' is the length of baffle section, in ft [the submerged length is $L'+15$ ft ($L'+4.6$ m) to allow for an inlet and exit from the baffle section].

3.6.9 Drain Systems

A drain system that is connected directly to pressure vessels is called a "pressure" or "closed" drain system. A drain system that collects liquids that spill on the ground is an "atmospheric," "gravity," or "open" drain. The liquid in a closed drain system must be assumed to contain dissolved gases that flash in the drain system and can become a hazard if not handled properly. In addition, it must be assumed that a closed drain valve could be left open by accident. Once the liquid has drained out of the vessel, a large amount of gas will flow out of the vessel into the closed drain system (gas blow-by) and will have to be handled safely. Thus, closed drain systems should always be routed to a pressure vessel and should never be connected to an open drain system.

Liquid gathered in an open drain system is typically rainwater or wash-down water contaminated with oil. With the oil usually circulated back into the process, every attempt should be made to minimize the amount of aerated water that is recycled with the oil. This goal is best achieved by routing open drains to a sump tank that has a gas blanket and operates as a skimmer. To keep gas from the skimmer from flowing out the drain, a water seal should be built into the inlet to the sump tank. Water seals should also be installed on laterals from separate buildings or enclosures to keep the open drain system from being a conduit of gas from one location in the facility to another.

3.7 Information Required for Design

3.7.1 Effluent Quality

- Allowable concentration of oil in effluent water. The U.S. Environmental Protection Agency (EPA) establishes the maximum amount of oil and grease content in water that can be discharged into navigable waters of the United States. In other locations local governments or governing bodies will establish this criterion.
- Produced water flow rate (Q_w, bwpd)
- Specific water gravity of produced water $(SG)_w$. Assume 1.07 if data are not available.
- Wastewater viscosity at flowing temperatures (μ, cp). Assume 1.0 cp if data are not available.
- Specific gravity of oil at flowing temperature $(SG)_o$
- Design rainfall rate (R_w, in./h). Assume 2 in./h in the Gulf of Mexico
- Flow rate for wash down (QWD, BPD). Assume 1500 bwpd per 50-gpm wash-down hose.
- Concentration of soluble oil at discharge conditions (mg/l or ppm)

3.7.2 Influent Water Quality

Produced Water

The first step in choosing a water treating system is to characterize the influent water streams. It is necessary to know both the *oil concentration* in this stream and the *particle size distribution* associated with this concentration. This is best determined from field samples and laboratory data.

Various attempts have been made to develop design procedures to determine oil concentration in water outlets from properly designed free-water knockouts and treaters. A conservative assumption would be that the water contains less than 1000–2000 mg/l of dispersed oil.

It is possible to theoretically trace the particle size distribution up the tubing, through the choke, flow lines, manifolds, and production equipment into the free-water knockout using equations presented in previous sections. However, many of the parameters needed to solve these equations, especially those involving coalescence, are unknown.

Because of the dispersion through the water dump valve, the oil size distribution at the outlet of a free-water knockout or heater-treater is not a significant design parameter. From the dispersion theory it can be shown that after passing through the dump valve a maximum droplet diameter on the order of 10–50 µm will exist no matter what the droplet size distribution was upstream of this valve.

If there were sufficient time for coalescence to occur in the piping downstream of the dump valve, then the maximum droplet diameter would be defined by Equation (3.2) prior to the water entering the first vessel in the water treating system.

The solution of this equation requires the determination of surface tension. The surface tension of an oil droplet in a water continuous phase is normally between 1 and 50 dyn/cm. It is not possible to predict the value without actual laboratory measurements in the produced water. Small amounts of impurities in the produced water can lower the surface tension significantly from what might be measured in synthetic water. In addition, as these impurities change with time, so will the surface tension. In the absence of data it is recommended that a maximum diameter of between 250 and 500 µm be used for design.

It is clear that there will be distribution of droplet sizes from zero to the maximum size, and this distribution will depend upon parameters unknown at the time of initial design. Experimental data indicate that a conservative assumption for design would be to characterize the distribution by a straight line, as shown in Figure 3.3.

Soluble Oil

In every system substances that show up as "oil" in the laboratory test procedure will be dissolved in the water. This is especially true where samples are acidized for "stabilization" prior to extraction with a solvent. This soluble oil cannot be removed by the systems discussed in this chapter. The soluble oil concentration should be subtracted from the discharge criteria to obtain a concentration of dispersed oil for design. Soluble oil concentrations as high as 1000 mg/l have been measured on rare occasions.

Deck Drainage

Federal regulations and most authorities having jurisdiction require that "free oil" be removed from deck drainage prior to disposal. It is extremely difficult to predict an oil drop size distribution for rainwater or wash-down water that is collected in an open drain system, and regulations do not define what size droplet is meant by "free oil."

Long-standing refinery practice is to size the drain water treating equipment to remove all oil droplets 150 μm in diameter or larger. If no other data are available, it is recommended that this be used in sizing sumps and disposal piles.

3.8 Equipment Selection Procedure

It is desirable to bring information included earlier into a format that can be used by the design engineer in selecting and sizing the individual pieces of equipment needed for a total water treating system. Federal regulations and most authorities having jurisdiction require that produced water from the free-water knockout receive at least some form of primary treatment before being sent to a disposal pile or skim pile. Deck drainage may be routed to a properly sized disposal pile that will remove "free oil."

Every water treating system design must begin with the sizing, for liquid separation of a free-water knockout, heater treater, or three-phase separator. These vessels should be sized in accordance with the procedures discussed in previous chapters.

With the exception of these restraints the design engineer is free to arrange the system as he or she sees fit. There are many potential combinations of the equipment previously described. Under a certain set of circumstances, it may be appropriate to dump the water from a free-water knockout directly to a skim tank for final treatment before discharge. Under other circumstances a full system of plate coalescers, flotation units, and skim piles may be needed. In the final analysis the choice of a particular combination of equipment and its sizing must rely rather heavily on the judgment and experience of the design engineer. The following procedure is meant only as a guideline and not as a substitute for this judgment and experience. Many of the correlations presented herein should be refined as new data and operating experience become available. In no instance is this procedure meant to be used without proper weight given to operational experience in the specific area.

1. Determine the oil content of the produced water influent. In the absence of other information, 1000–2000 mg/l could be assumed.
2. Determine the dispersed oil effluent quality. In the absence of other information, use 23 mg/l for design in the Gulf of Mexico and other similar areas (29 mg/l allowed less than 6 mg/l dissolved oil).

3. Determine oil drop size distribution in the influent produced water stream. Use a straight-line distribution with a maximum diameter of 250–500 μm in the absence of better data.

4. Determine the oil particle diameter that must be treated to meet effluent quality required. This can be calculated as effluent quality divided by influent quality times the maximum oil particle diameter calculated in step 3.

5. If there is a large amount of space available (as in an onshore location), consider an SP Pack system. Proceed to step 10. If the answer to step 4 is less than 30–50 μm, a flotation unit or cyclone is needed. Proceed to step 6. If the answer to step 4 is greater than 30 μm, a skim tank or plate coalescer could be used as a single stage of treatment, but this is not really recommended. Proceed to step 9.

6. Determine flotation cell influent quality from the required effluent quality assuming 90% removal. Influent quality is effluent quality desired times 10.

7. If required flotation cell influent quality is less than the quality determined in step 1, determine the particle diameter that must be treated in skim tank or plate coalescers to meet this quality. This can be calculated as the flotation cell influent quality divided by the influent quality determined in step 1 times the maximum particle diameter calculated in step 3.

8. Determine effluent from hydrocyclone, assuming that it is 90% efficient, and determine the particle diameter that must be treated in the downstream skim vessel, assuming that $d_{max} = 500$. This value can be calculated as 500 times the dispersed oil effluent quality (step 1) divided by the effluent concentration from the hydrocyclone.

9. Determine skimmer dimensions.
   a. Choose horizontal or vertical configuration.
   b. Choose pressure vessel or atmospheric vessel.
   c. Determine size. Refer to appropriate equations.

10. Determine overall efficiency required, efficiency per stage, and number of stages for an SP Pack system.

11. Determine plate coalescers' dimensions.
    a. Choose CPI or cross-flow configuration.
    b. Determine size. Refer to appropriate equations.

12. Choose skim tank, SP Pack, or plate coalescers for application, considering cost and space available.

13. Choose method of handling deck drainage.
    a. Determine whether rainwater rate or wash-down rate governs design.
    b. Size disposal pile for drainage assuming 150-μm drop removal. Refer to appropriate equations.
    c. If disposal pile diameter is too large,

i. Size sump tank to use with disposal pile (refer to appropriate equations), or

ii. Size skim pile (refer to appropriate equations).

3.9 Equipment Specification

Once the equipment types are selected using the previous procedure, the design equations presented in this chapter can be used to specify the main size parameters for each of the equipment types.

3.9.1 Skim Tank

1. *Horizontal vessel designs.* The internal diameter and seam-to-seam length of the vessel can be determined. The effective length of the vessel can be assumed to be 75% of the seam-to-seam length.
2. *Vertical vessel designs.* The internal diameter and height of the water column can be determined. The vessel height can be determined by adding approximately 3 ft to the water column height.

3.9.2 SP Pack System

The number and size of tanks can be determined. Alternatively, the dimensions and number of compartments in a horizontal flume can be specified.

3.9.3 CPI Separator

The number of plate packs can be determined.

3.9.4 Cross-Flow Devices

The acceptable dimensions of the plate pack area can be determined. The actual dimensions depend on the manufacturers' standard sizes.

3.9.5 Flotation Cell

Information is given to select a size from the manufacturers' data.

3.9.6 Disposal Pile

The internal diameter and length can be determined. For a skim pile the length of the baffle section can be chosen.

Example 3.2: Design the Produced Water Treating System for the Data Given

Given:
   40 °API
   5000 bwpd (33 m³/h)
   Deck size is 2500 ft² (232.3 m²)
   48 mg/l discharge criteria (48 mg/l)
   Water gravity-feeds to system

   Step 1. Assume 6 mg/l soluble oil, and oil concentration in produced water is 1000 mg/l.
   Step 2. Effluent quality required is 48 mg/l. Assume 6 mg/l dissolved oil. Therefore, effluent quality required is 42 mg/l.
   Step 3. Assume maximum diameter of oil particle $(d_{max}) = 500$ μm.
   Step 4. Using Figure 3.3, the size of oil droplet that must be removed to reduce the oil concentration from 1000 to 42 mg/l is

$$\frac{d_m}{500} = \frac{42}{1000},$$

$$d_m = 21 \text{ μm.}$$

   Step 5. Consider an SP series tank treating system. See step 10. If SP Packs are not used, since $d_m < 30$ μm, a flotation unit or hydrocyclone must be used. Proceed to step 6. (Note: since d_m is close to 30 μm, it may be possible to treat this water without a flotation unit. We will take the more conservative case for this example.)
   Step 6. Since the flotation cell is 90% efficient, in order to meet the design requirements of 42 mg/l it will be necessary to have an influent quality of 420 mg/l. This is lower than the 1000-mg/l concentration in the produced water assumed in step 1. Therefore, it is necessary to install a primary treating device upstream of the flotation unit.
   Step 7. Using Figure 3.52, the size of oil droplet that must be removed to reduce the oil concentration from 1000 to 420 mg/l is

$$\frac{d_m}{500} = \frac{420}{1000},$$

$$d_m = 210 \text{ μm.}$$

   Step 8. Inlet to water treating system is at too low a pressure for a hydrocyclone. Size a skim vessel upstream of the flotation unit.
   Step 9. Skim vessel design. Pressure vessel is needed for process considerations (e.g., fluid flow, gas blow-by).
   a. Assume horizontal pressure vessel.

Settling equation

Field units

$$dL_{\text{eff}} = \frac{1000 Q_w \mu_w}{(\Delta SG)(d_m)^2},$$

$\mu_w = 1.0$ (assumed),
$(SG)_w = 1.07$ (assumed),
$(SG)_o = 0.83$ (calculated),

$$dL_{\text{eff}} \frac{(1000)(5000)(1.0)}{(0.24)(210)^2} = 472.$$

Assume various diameters (d) and solve for L_{eff}.

d *(in.)*	L_{eff} *(ft)*	*Actual Length (ft)*
24	19.7	26.3
48	9.8	13.1
60	7.9	10.5

Retention time equation

Assume retention time of 10 min.

$$d^2 L_{\text{eff}} = 1.4(t_r)_w Q_w,$$

$$d^2 L_{\text{eff}} = (1.4)(10)(5000) = 7000.$$

d *(in.)*	L_{eff} *(ft)*	*Actual Length (ft)*
48	30.4	40.4
72	13.5	17.9
84	9.9	13.1
96	7.6	10.1

SI units

$$dL_{\text{eff}} = \frac{1,145,734 Q_w \mu_w}{(\Delta SG)(d_m)^2},$$

$\mu_w = 1.0$ (assumed),
$(SG)_w = 1.07$ (assumed),
$(SG)_o = 0.83$ (calculated),

$$dL_{\text{eff}} = \frac{(1,145,734)(33)(1.0)}{(0.24)(210)^2} = 3572.$$

Assume various diameters (d) and solve for L_{eff}.

d *(mm)*	L_{eff} *(m)*	*Actual Length (m)*
609.6	5.9	8.0
1219.2	3.0	4.0
1542	2.3	3.2

Retention time equation

Assume retention time of 10 min.

$$d^2 L_{eff} = 42{,}441 (t_r)_w Q_w,$$
$$d^2 L_{eff} = (42{,}441)(10)(33).$$

d *(mm)*	L_{eff} *(m)*	*Actual Length (m)*
1219	9.4	12.3
1829	4.2	5.5
2134	3.1	4.0
2438	2.4	3.1

b. Assume vertical pressure vessel.

Field units

Settling equation

$$d^2 = 6691 F \frac{Q_w \mu_w}{(\Delta SG)(d_m)^2},$$

$$F = 1.0 \text{ assumed,}$$

$$d^2 = \frac{(6691)(1.0)(5000)(1.0)}{(0.24)(210)^2},$$

$$d = 56.22 \text{ in.}$$

Retention time equation

$$H = 0.7 \frac{(t_r)_w Q_w}{d^2} = 0.7 \frac{(10)(5000)}{d^2}$$

d (in.)	L_{eff} (ft)	Seam-to-Seam Height (ft)
60	9.72	12.7
66	8.03	11.0
72	6.75	9.8

SI units

Settling equation

$$d^2 = 6365 \times 10^8 \frac{Q_w \mu_w}{(\Delta SG)(d_m)^2},$$

$$F = 1.0 \text{ assumed,}$$

$$d^2 = \frac{(6365 \times 10^8)(1.0)(33)(1.0)}{(0.24)(210)^2},$$

$$d = 1409 \text{ mm.}$$

Retention time equation

$$H = \frac{21{,}218(t_r)_w Q_w}{d^2} = 21{,}218 \frac{(10)(33)}{d^2}$$

d (mm)	L_{eff} (m)	Seam-to-Seam Height (m)
1524	3.0	3.9
1676.4	2.5	3.4
1829	2.1	3.0

A vertical vessel 60 in. (1524 mm)×12.5 ft (3.8 m) or 72 in. (1829 mm)×10 ft (3 m) would satisfy all the parameters. Depending on cost and space considerations, we recommend a 72-in. (1829-mm)×10-ft (3-m) vertical skimmer vessel for this application.

Step 10. Investigate SP Packs in tanks as an option. Calculate overall efficiency required:

Field units

$$E_t = \frac{1000 - 42}{1000} = 0.958.$$

Assume 10-ft (3-m)- diameter vertical tanks with $F = 2$:

$$d_m^2 = 6691F\frac{Q_w\mu_w}{(\Delta SG)(d^2)},$$

$$d_m^2 = \frac{6691(2)(5000)(1.0)}{(0.24)(120)^2},$$

$$d_m = 139.$$

Assume SP Pack grows 1000-µm drops:

$$E = 1 - \frac{d_m}{1000} = 0.861.$$

One acceptable choice is two 10-ft (3-m)- diameter SP tanks in series.

$$E_t = 1 - (1 - 0.861)^2 = 0.981$$

SI units

$$E_t = \frac{1000 - 42}{1000} = 0.958$$

Assume 10-ft (3-m)- diameter vertical tanks with $F = 2$:

$$d_m^2 = 6365 \times 10^8 F\frac{Q_w\mu_w}{(\Delta SG)(d^2)},$$

$$d_m^2 = \frac{6365 \times 10^8(2)(33)(1.0)}{(0.24)(3000)^2},$$

$$d_m = 139 \text{ µm}.$$

Assume SP Pack grows 1000-µm drops:

$$E = 1 - \frac{d_m}{1000} = 0.861.$$

One acceptable choice is two 10-ft (3-m)- diameter SP tanks in series.

$$E_t = 1 - (1 - 0.861)^2 = 0.981$$

Step 11. Check for alternate selection of CPI.

Field units

$$\text{Number of packs} = 0.077\frac{Q_w\mu_w}{(\Delta SG)d_m^2}$$

$$= \frac{(0.077)(5000)(1.0)}{(0.24)(210)^2}$$

$$= 0.04 \text{ packs},$$

$$Q_w < 20{,}000 \text{: use 1 pack CPI}.$$

SI units

$$\text{Number of packs} = 11.67 \frac{Q_w \mu_w}{(\Delta SG) d_m^2}$$

$$= \frac{(11.67)(33)(1.0)}{(0.24)(210)^2}$$

$$= 0.04 \text{ packs,}$$

$$Q_w < 132\text{: use 1 pack CPI.}$$

Step 12. Recommended skimmer vessel over CPI as skimmer will take up about same space, will cost less, and will not be susceptible to plugging. Note that it would also be possible to investigate other configurations such as skim vessel, SP Pack, CPI, etc. as alternatives to the use of a flotation unit.

Step 13. Sump design. Sump is to be designed to handle the maximum of either rainwater or wash-down hose rate.

c. Rainwater rate:

Field units

Assume

$$R_w = \text{rainfall rate; 2 in./h, } A_D = \text{deck area; 2500 ft}^2,$$
$$Q_w = 0.356 A_D R_w = (0.356)(2500)(2) = 1780 \text{ bwpd.}$$

SI units

Assume

$$R_w = 50.8 \text{mm/h}, \quad A_D = 232.3 \text{m}^2,$$
$$Q_w = 0.001 \ A_D R_w = (0.001)(232.3)(50.8) = 11.8 \text{m}^3/\text{h.}$$

d. Wash-down rate:

Field units

Assume

$$N = 2,$$

$$Q_{WD} = 1500N$$

$$= 1500(2)$$

$$= 3000 \text{ bwpd.}$$

Assume

$$N = 2,$$

$$Q_{WD} = 9.92 \, N$$

$$= 9.92(2)$$

$$= 19.84 \text{m}^3/\text{h}.$$

The minimum design usually calls for two hoses.
Because freshwater enters the sump via the drains, the sump tank must be sized using a specific gravity of 1.0 and a viscosity of 1.0 for freshwater.

e. Assume horizontal rectangular cross-section sump.

Settling equation

Field units

$$WL_{\text{eff}} = 70 \frac{Q_w \mu_w}{(\Delta SG) d_m^2},$$

$$WL_{\text{eff}} = 70 \frac{(3000)(1.0)}{(0.150)(150)^2},$$

$$WL_{\text{eff}} = 62.2,$$

$$W = \text{width, ft.}$$

L_{eff} = effective length in which separation occurs, ft
H = height of tank, which is 1.5 times higher than water level within tank, or $0.75W$.

Tank Width (ft)	Tank L_{eff} (ft)	Seam-to-Seam Length	Height (ft)
4	15.6	20.7	3.0
5	12.4	16.2	3.8
6	10.4	13.5	4.5

SI units

$$WL_{\text{eff}} = 950 \frac{Q_w \mu_w}{(\Delta SG) d_m^2},$$

$$WL_{\text{eff}} = 950 \frac{(19.84)(1.0)}{(0.150)(150)^2},$$

$WL_{eff} = 5.6,$
W = width, m,
L_{eff} = effective length in which separation occurs, m,
H = height of tank, which is 1.5 times higher than water
       level within tank, or 0.75W.

Tank Width (m)	Tank L_{eff} (m)	Seam-to-Seam	Height (m)
1.2	4.7	6.2	0.9
1.5	3.7	4.8	1.1
1.8	3.1	4.1	1.4

A horizontal tank 6 (1.83 m)×14 (4.3 m)×5 (1.52 m) would satisfy all design parameters.

f. If it is determined that the dimensions of the sump tank are inappropriately large for the platform, an SP Pack can be added upstream of the sump tank to increase oil droplet size by approximately two times the inlet droplet size.

Therefore, the sump tank size with an SP Pack can be determined by

Field units

$$WL_{eff} = \frac{70(3000)(1.0)}{(0.15)(300)^2} = 15.6,$$

W = width, ft (m),
L_{eff} = effective length in which separation occurs,
       ft (m),
H = height of tank, which is 1.5 times higher
       than water level within tank, or 0.75W.

Tank Width (ft)	Tank L_{eff} (ft)	Seam-to-Seam Length (ft)	Height (ft)
3	5.2	6.9	2.3
4	3.9	5.2	3.0
5	3.1	4.1	3.8

SI units

$$WL_{eff} = \frac{950(19.84)(1.0)}{(0.15)(300)^2} = 1.4,$$

W = width, ft (m),
L_{eff} = effective length in which separation occurs,
       ft (m),
H = height of tank, which is 1.5 times higher
       than water level within tank, or 0.75W.

Tank Width (m)	Tank L$_{eff}$ (m)	Seam-to-Seam (m)	Height (m)
0.9	1.6	2.1	0.68
1.2	1.2	1.6	0.9
1.5	0.9	1.2	1.13

A horizontal tank (with an SP Pack) 4 ft (1.2 m) × 4 ft (1.2 m) × 5 ft (1.5 m) would satisfy all design parameters. It can be seen that by adding an SP Pack, sump tank sizes can be substantially reduced.

References

Bradley, H. B., and Collins, A. G. Properties of produced waters, *Petroleum Engineering Handbook*, SPE, Richardson, TX, 1987.

Schramn, L. L. "Basic Principles", *Emulsion Fundamentals, and Applications in the Petroleum Industry*, Schramn L. L.(Ed.), American Chemical Society, Washington, DC, 1992.

Callaghan, D., and Baumgartner, W. *Characterization of Residual Hydrocarbons in Produced Water Discharged from Gas Production Platforms*, SPE 20881, 1990.

Jacobs, R. P. W. M., Grant, R. O. H., Kwant, J., Marquenie, J. M., and Mentzer, E. "The composition of produced water from shell operated oil and gas production in the North Sea", Ray, J. P., and *Produced Water: Technological/ Environmental Issues and Solutions*, Englehart, R. (Eds.), Plenum Press, New York, 1992.

Jackson, G. F., Hume, E., Wade, M. J., and Kirsch, M. *Oil Content in Produced Brine of Ten Louisiana Production Platforms*, Gulf Publishing Company, Houston, TX, 1986.

Patton, C. C. *Applied Water Technology*, Campbell Petroleum Series, 1986.

Arnold, K. E., and Stewart, M. I. *Surface Production Operations-Design of Oil Handling Facilities*, 3rd Edition, Elsevier, 2008.

CHAPTER 4

Water Injection Systems

4.1 Introduction

Oil-field waters usually contain impurities. These impurities are classified as dissolved minerals, dissolved gases, or suspended solids. Suspended solids can be naturally occurring, generated by precipitation of dissolved solids, generated as products of corrosion, or created by microbiological activity. Changes in temperature, pressure, pH, or the mixing of waters from different sources may cause *scaling*, which is precipitation of dissolved solids. Suspended solids may settle out of the water stream or may be carried as a suspension in flowing water.

The two primary sources of freshwater are surface water and groundwater. A portion of the rain or melting snow and ice at the earth's surface soaks into the ground, while part of it collects in ponds and lakes or runs off into creeks and rivers. This latter portion is termed "surface water."

Water encountered in production operations usually comes from separated produced water or from wells specifically drilling into a subsurface water aquifer. The latter is often called "source water," is often brackish, and may contain a large quantity of dissolved solids.

This chapter provides information about equipment selection and sizing for removing suspended solids and dissolved gases from water. The water's source affects the types and amounts of contaminants in the water. For example, produced water will be contaminated by some hydrocarbons.

The treatment of water to remove calcium and magnesium dissolved solids ("water softening") is important, especially if the water is to be used as boiler feed water for the generation of steam, as in a steam flood. Nevertheless, a discussion of processes and equipment for water softening and removing other dissolved solids is beyond the scope of this chapter.

The removal of suspended solids and dissolved gases from water may be desirable for a variety of reasons, the most common of which are to prepare the water for injection into a producing formation and to minimize the corrosion and solids build-up in surface equipment. Prior to injecting water, it may be important to remove solids above a certain size to minimize damage to the formation caused by solids plugging. This plugging can limit injection volumes, increase pump horsepower requirements, or lead to fracturing of the reservoir rock. Dissolved gases such as oxygen in the water may promote bacteria growth within the formation, or they may speed the process of corrosion.

The presence of oxygen or hydrogen sulfide (H_2S) in water can lead to the formation of FeS, Fe_2O_3, elemental sulfur particles, and scale. These solid particles may form after the water is already downstream of solid removal equipment. Without proper consideration of dissolved gases, the benefits of installing solids removal equipment can thus be partially negated.

In any solids removal system, there is a need for equipment to handle the bulk solids or sludge removed from the water and a procedure for removal of these solids. For many common water injection systems, the amount of bulk solids to be removed can be rather large. If the solids are free of oil, they may be disposed of in slurry piped to pits onshore or overboard offshore. If coated with oil, they may require treating prior to disposal. Treating oil from solids is beyond the scope of this chapter. Oil is normally separated from produced solids by abrasion in hydrocyclones or by washing with detergents or solvents.

Selection of a specific design of water treating system for removing suspended solids and dissolved gases from a water source requires establishing the year-round quality of the water source. This determination normally requires that tests be performed to identify the amount of dissolved gases [primarily oxygen and hydrogen sulfide (H_2S)] present in the water, the total mass of suspended solids and their particle size distribution, and the amount of oil present in the source. In addition, if a source of water is to be injected into a reservoir, it must be checked to ensure that it is compatible with reservoir water; that is, that under reservoir conditions, dissolved solids will not precipitate in sufficient quantity to plug the well or reservoir. Similarly, if two sources of water are to be mixed on the surface, they must be checked for compatibility under surface conditions of pressure, temperature, and pH. The tests are normally performed by laboratories that specialize in offering these services. (Determination of allowable concentration and particle size of solids and the acceptable level of dissolved gases in injection water is beyond the scope of this text.)

First, the theory involved in the various processes for removal of solids and dissolved gases from water is discussed. Next, the equipment used in both processes is discussed, and, finally, a design procedure to follow in selecting the equipment for a specific application is presented.

Treating water for solids removal and for removal of dissolved gases are really two separate concepts, using two separate sets of theories and equipment. They are combined in this chapter only because both are usually considered together when designers plan a water treating system for water injection. With this presentation the designer can select the design required to prepare any water stream for several common uses. Water softening, potable water making, and boiler feed water preparation, however, are several important water treating topics that are not within the scope of this chapter.

4.2 Solids Removal Theory

4.2.1 Removal of Suspended Solids from Water

For a variety of reasons, it may be desirable to remove suspended solids from a water stream. This removal is most commonly done as part of a water injection system for water-flood or enhanced oil recovery. It may also be necessary to remove suspended solids prior to injecting produced water in disposal wells.

Two different principles have been used to develop equipment for removing suspended solids from water. Gravity settling uses the density difference between the solid particle and the water to remove the solids; filtration traps the solids within a filter medium that allows water to pass.

The quantity of suspended solids in a water stream is normally expressed in milligrams per liter (mg/l) or parts per million (ppm) by weight (mg/l divided by water specific gravity equals ppm). The size of the suspended particles is usually expressed as a diameter stated in units of micrometers (10^{-6} m), also called microns. The capability of the equipment or filters to remove suspended solids is expressed in terms of removal of a percentage of all suspended solids having a diameter greater than a specified micron size. These values will usually range from 150 μm for gravity separators to less than 0.5 μm for filters. Suspended solids less than 40 μm in diameter cannot be seen with the naked eye. Figure 4.1 shows relative sizes for a variety of common materials.

4.2.2 Gravity Settling

The force of gravity may be used to remove solid particles from water if the density of these particles is not the same as the density of the water. Typically, solid particles have a density greater than water; therefore, they fall relative to the water due to the force of gravity. The terminal settling velocity is such that the gravitational force on the particle equals the drag force resisting its motion due to friction.

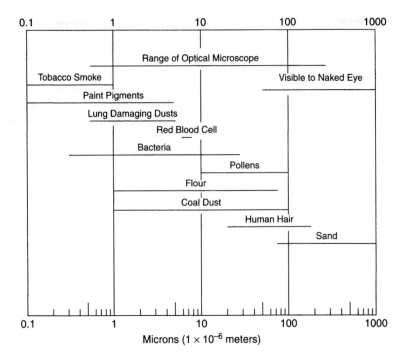

FIGURE 4.1. Relative sizes of common materials.

Assuming the particle is roughly spherical, the drag force may be determined as follows:

$$F_D = C_D A_\rho \left[\frac{V_t^2}{2_g}\right] \qquad (4.1)$$

where

F_D = drag force, lb (kg),
C_D = drag coefficient,
A = cross-sectional area of particle, ft^2 (m^2),
ρ = density of the continuous phase, lb/ft^3 (kg/m^3),
V_t = terminal settling velocity of the particle, ft/s (m/sec),
G = gravitational constant, 32.2 ft/sec^2 (9.81 m/sec^2).

The terminal settling velocity of small particles through water is low, and flow around the particle is laminar. Therefore, Stokes' law may be applied to determine the drag coefficient as follows:

$$C_D = \frac{24}{Re}, \qquad (4.2)$$

where Re is the Reynolds number.

By equating the gravitationally induced negative buoyant force with the drag force, one may derive the following equation for calculating the terminal settling velocity of the particle:

Fields Units

$$V_t = \frac{1.78 \times 10^{-6}(\Delta SG)d_m^2}{\mu},$$ (4.3a)

SI Units

$$V_t = 5.44 \times 10^{-10} \frac{(\Delta SG)d_m^2}{\mu},$$ (4.3b)

where

ΔSG = difference in specific gravity of the particle and the water,
d_m = particle diameter, µm,
μ = viscosity of the water, cp (Pas).

Equations (4.3a) and (4.3b) may be used to size any of several types of equipment designed to use gravity settling. Such devices as settling ponds, pits, flumes, and tanks are commonly used onshore where space is available.

Parallel plate interceptors, such as CPIs (corrugated plate interceptors) and cross-flow separators, can be effective at removing suspended solids from water. However, the solids tend to cling to the plates and plug the plate pack. For this reason, parallel plate interceptors are not normally used to remove large quantities of solids from water.

Other devices such as hydrocyclones and centrifuges also take advantage of the density differences between the water and the suspended solid particles. These devices induce centrifugal forces in the water, causing the heavy solid particles to move away from the axis of rotation. Gravity settling relies on low velocities and large particle sizes (greater than 10 µm) to be most effective.

4.2.3 Flotation Units

Fines, oil, and oil wetted solids, which cannot be removed by gravity settling, may be removed by gas flotation units. In these units, a bubble attaches to the contaminant particle, lowering the effective weight of the particle and allowing it to rise to the surface, where it is removed by skimming.

Flotation units can be classified as dissolved gas or induced gas, depending on the gas supply mechanism. Dissolved gas flotation (DGF) introduces a gas/water solution in saturation at high pressure into the wastewater stream. Induced gas flotation (IGF) forms gas bubbles and provides turbulence for mixing by rotating mechanical diffusers or by recirculating a portion of the water through gas eductors.

Although flotation is a very common process used in the mining industry to separate metals from crushed rock slurries, it is not commonly used for solids removal in production operations.

4.2.4 Filtration

Filtration can be used to remove suspended solid particles from water by passing the water through a porous filter medium. As the water passes through the small pores in the filter medium, particles larger than the pores become trapped. The size of the pores in a filter medium determines the smallest particles that may be trapped.

Suspended solids are separated from fluids via three mechanisms: inertial impaction, differential interception, and direct interception.

Inertial Impaction
Particles (1–10 μm) in a fluid stream have mass and velocity and, hence, have a momentum associated with them. As the liquid and entrained particles pass through a filter media, the liquid stream will take the path of least resistance to flow and will be diverted around the fiber. The particles, because of their momentum, tend to travel in a straight line and, as a result, those particles located at or near the center of the flow line will strike or impact upon the fiber and be removed. Figure 4.2 illustrates this process. The fluid stream, shown as solid lines, flows around the filter fibers while the particles continue along their path, shown as dashed lines, and strike the fibers. Generally, larger particles will more readily deviate from the flow lines than small ones. In practice, however, because the differential densities of the particles and fluids are very small, deviation from the liquid flow line is much less and hence inertial impaction in liquid filtration plays a relatively small role.

Diffusional Interception
For particles that are extremely small (i.e., those with very little mass and less than 0.3 μm in diameter), separation can result from diffusional interception. In this process, particles are in collision with the liquid molecules. These frequent collisions cause the suspended particles to move in a random fashion around the fluid flow lines. Such movement, which can be observed microscopically, is called "Brownian motion." Brownian motion causes these smaller particles to deviate from the fluid flow lines and hence increase the likelihood

FLUID FLOW STREAMLINES

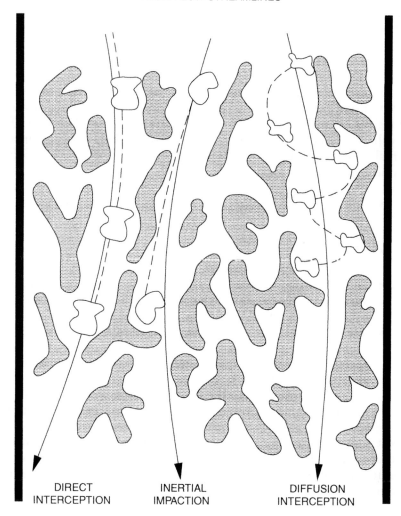

FIGURE 4.2. Filtration mechanisms.

of their striking the fiber surface and being removed. Figure 4.2 shows the particle flow characterized by Brownian motion and impacting the filter fibers. As with inertial impaction, diffusional interception has a minor role in liquid filtration because of the inherent nature of liquid flow, which tends to reduce the lateral movement or excursions of the particle away from the fluid flow lines.

Direct Interception
While inertial impaction and diffusional interception are not as effective in liquid service as in gas service, direct interception is equally as

effective in both and is the desired mechanism for separating particles from liquids. In a filter medium, one observes not a single fiber, but rather an assembly of a large number of such fibers. These fibers define openings through which the fluid passes. If the particles in the fluid are larger than the pores or openings in the filter medium, they will be removed as a result of direct interception. Figure 4.2 portrays this removal mechanism. Direct interception is easily understood in the case of a woven mesh filter with uniform pores and no thickness or depth; once a particle passes through an opening, it proceeds unhindered downstream. Yet such a filter will collect a very significant proportion of particles whose diameter is smaller than the openings or pores of the medium. Several factors that help account for this collection are:

- Most suspended particles, even if quite small when viewed from some directions, are irregular in shape and hence can "bridge" an opening.
- A bridging effect can also occur if two or more particles strike an opening simultaneously.
- Once a particle has been stopped by a pore, that pore is at least partially occluded and subsequently will be able to separate even smaller particles from the liquid stream.
- Specific surface interactions can cause a small particle to adhere to the surface of the internal pores of the medium. For example, a particle considerably smaller than a pore is likely to adhere to that pore provided the two surfaces are oppositely charged. A very strong, negatively charged filter can cause a positive charge to be induced on a less strongly charged negative particle.

Direct interception can also obviously occur in filters in which the pore openings are not uniform, but instead vary in size (but within carefully controlled limits) throughout the thickness of the filter medium, resulting in a tortuous flow path.

4.3 Filter Types

In recent years it has become increasingly common to classify filters and filter media as either "depth type" or "surface type." Unfortunately, filter manufacturers have been unable to agree upon an "official" definition of the terms. As a result, much misunderstanding is encountered in the field on this subject. Hopefully, this discussion will separate fact from fiction.

A number of available filter media provide different pore sizes for solids removal. Depending on their construction, filter media may be divided into either nonfixed-pore or fixed-pore types.

4.3.1 Nonfixed-Pore Structure Media

Nonfixed-pore structure filters depend principally on the filtration mechanisms of inertial impaction and/or diffusional interception to trap particles within the spaces of their internal structure.

The nonfixed filters are constructed of nonrigid media. Variations in pressure drop through nonfixed filters may cause minor deformation or movement of the filter medium, potentially changing the size of some of the pores in the medium (hence the name "nonfixed pore"). Nonfixed-pore filters are by far the most common type of filters and include the following:

- Unbonded fiberglass cartridges,
- Cotton-wound or sock filters,
- Molded cellulose cartridges,
- Spun-wound polypropylene cartridges,
- Sand and other granular media beds,
- Diatomaceous earth filters.

Nonfixed-pore structure-type filters depend not only on trapping but also on adsorption to retain particles. As long as the dislodging force exerted by the fluid is less than the force retaining the particle, the particle will remain attached to the medium. However, when such a filter has been on-stream for a length of time and has collected a certain amount of particulate matter, a sudden increase in flow and/or pressure can overcome these retentive forces and cause the release downstream of some of the particles. This unloading will frequently occur after the filter has been in use for some time and can give a false impression of long service life for the filter.

Most nonfixed-pore structure filters are subject to media migration. This means that parts of the filter medium become detached and continue to pass downstream, contaminating the effluent (fluid that has passed through the filter). Media migration is also sometimes incorrectly taken to include release of "built-in" contamination—for example, dust and fibers picked up by the filter during its manufacture.

4.3.2 Fixed-Pore Structure Media

Fixed-pore media filters consist of either layers of medium or a single layer of medium having depth, depend heavily on the mechanism of direct interception to do their job, and are so constructed that the structural portions of the medium cannot distort and the flow path through the medium is tortuous. It is true that such filters retain some particles by adsorption as a result inertial impaction and diffusional interception. It is also true that they contain pores larger than their removal rating. However, pore size is controlled in manufacture so that quantitative removal of particles larger than a given size can be assured.

Fixed-pore filters are constructed such that the pore size does not change. Such filters represent relatively new technology in filter medium construction for oil-field use and include the following:

- resin-impregnated cellulose cartridges,
- resin-bonded glass fiber cartridges, and
- continuous polypropylene cartridges.

As solids are trapped in a filter, some of the available flow paths are blocked. This blockage causes an increased pressure drop through the filter and may cause minor movement within the filter medium, which can result in unloading and/or media migration. (Unloading refers to previously trapped solids being released downstream; media migration refers to portions of the media being released downstream.) Media migration almost always has some unloading associated with the release of media material. These two phenomena usually cause a sudden decrease in the pressure drop through the filter. It should be noted that, by definition, a fixed-pore filter does not exhibit unloading or media migration.

Eventually, a filter collects solids until the pressure drop is too large for continued operation. At this point, the filter medium must be replaced or cleaned. The amount of solids a filter may remove per unit volume is referred to as the filter's "solids loading." Different types of filters may have vastly different solids loading capabilities.

A filter's solids loading capacity is affected by the filter design and the particular medium used. Figure 4.3, for example, shows portions of three filters using different fiber media, which can be glass fiber, cellulose, cotton, or polypropylene. All three sections represent the same filter area with the same pore sizes. The only difference among the three is the fiber diameter used to form the medium. The

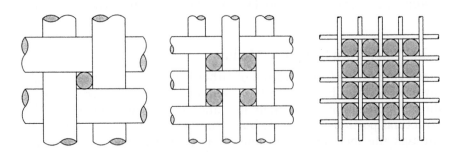

FIGURE 4.3. Fiber diameter affects filter's solids capacity.

right side represents a filter with 16 times as many pores per unit volume as the left filter; its solids loading should thus be much larger. This is true even though the filters may be made of the same materials and may appear the same to the naked eye.

4.3.3 Surface Media

A surface or screen filter is one in which all pores rest on a single plane, which therefore depends largely upon direct interception to separate particles from a fluid. Only a few filters on the market today, for example, woven wire mesh, woven cloth, and certain membrane filters, qualify as surface filters.

A surface or screen filter will stop all particles larger than the largest pore opening. While particles smaller than the largest pore may be stopped because of factors previously discussed (bridging, etc.), there is no guarantee that such particles will not pass downstream. Woven wire mesh filters are currently available with openings down to 5 μm.

4.3.4 Summary of Filter Types

It should be apparent that classifying filters as depth or surface is meaningless. Almost all filters exhibit "depth" when viewed under a microscope. A more meaningful classification of filters is as follows:

1. Nonfixed-pore structures have pores whose dimensions increase at high pressure ("wound," low-density filters).
2. Fixed-pore structures have pores that do not increase in size at high pressures (most membrane filters).
3. Screen media (woven cloth or screens).

Fixed-pore structure filters are superior for most purposes when compared with the screen type. They combine high dirt capacity per unit area with both absolute removal of particles larger than a given size and minimum release of collected particles smaller than this rated size under impulse conditions.

Nonfixed-pore structure filters do not have absolute ratings, are subject to media migration, and unload particles very badly on impulse. A comparison of dirt capacity between fixed-pore and nonfixed-pore depth filters is meaningless because the nominal removal ratings of nonfixed-pore structure filters bear little relationship to their behavior in service.

Upon reviewing the above information, it should become apparent that it matters little whether a filter is surface type or depth type. Then, how should one judge filters? Only two questions should concern the filter user. First, is the filter subject to unloading, channeling, or media migration? If so, then for most applications it's probably unwise to use the filter. If not, then look to the second question; can the manufacturer guarantee that the filter will reliably remove all particles that should be removed from the fluid in question and that the filter is safe to use?

4.4 Removal Ratings

Particular attention should be paid to filter rating. Various rating systems have evolved to describe the filtration capabilities of filter elements. Unfortunately, there is no generally accepted rating system, and this tends to confuse the filter user. Several of the rating systems now in use are described below.

4.4.1 Nominal Rating

Many filter manufacturers rely on a nominal filter rating, which has been defined by the National Fluid Power Association (NFPA). The NFPA states, "An arbitrary micron value assigned by the filter manufacturer, based upon removal of some percentage of all particles of a given size or larger. It is rarely well defined and not reproducible." In practice, a "containment" is introduced upstream of the filter element, and subsequently the effluent flow (flow downstream of the filter) is analyzed microscopically. A given nominal rating of a filter means that 98% by weight of the contaminant above the specialized size has been removed; 2% by weight of the contaminant has passed downstream.

Note that this is a gravimetric test rather than a particle count test. Counting particles upstream and downstream is a more meaningful way to measure filter effectiveness. The various tests used to give nonfixed-pore structure filters a nominal rating yield results that are misleading. Typical problems are as follows:

1. The 98% contaminant removal by weight is determined by using a specific containment at a given concentration and flow. If any one of the test conditions is changed, the test results could be altered significantly.
2. The 2% of the contaminant passing through the filter is not defined by the test. It is not uncommon for a filter with a nominal rating of 10 μm to pass particles downstream ranging in size from 30 to over 100 μm.
3. Test data are often not reproducible, particularly among different laboratories.

4. Some manufacturers do not base their nominal rating on 98% contaminant removal by weight, but instead a contamination removal efficiency of 95%, 90%, or even lower. Thus, it often happens that a filter with an absolute rating of 10 μm is actually finer than another filter with a nominal rating of 5 μm. Therefore, it is always advisable to check the criteria upon which a nominal rated filter is based.
5. The very high upstream contaminant concentrations used for such tests are not typical of normal system conditions and produce misleading high-efficiency values. It is common for a wire-mesh filter medium with a mean (average) pore size of 15 μm to pass a 10-μm nominal specification. However, at normal system contaminations, this same filter medium will pass almost all 10-μm particles.

Therefore, one cannot assume that a filter with a nominal rating of 10 μm will retain all or most particles 10 μm or larger. Yet some filter manufacturers continue to use only a nominal rating both because it makes their filters seem finer than they actually are and because it is impossible to place an absolute rating on a nonfixed-pore structure.

4.4.2 Absolute Rating

The NFPA defines an absolute rating as follows: "The diameter of the largest hard spherical particle that will pass through a filter under specified test conditions. It is an indication of the largest opening in the filter element." Such a rating can be assigned only to an integrally bonded medium.

There are several recognized tests for establishing the absolute rating of a filter. What test is used will depend on the manufacturer, on the type of medium to be tested, or sometimes on the processing industry. In all cases the filters have been rated by a "challenge" system. A filter is challenged by pumping through a suspension of a readily recognized contaminant (e.g., glass beads or a bacterial suspension) and both the influent and effluent examined for the presence of the test contaminant.

The challenge tests are destructive tests—that is, the challenged filter cannot be used after the test. Consequently, integrity tests for filters have been established, which are nondestructive and correlate with the destructive challenge test. In other words, if the test filter was successfully integrity tested by the nondestructive test, that would mean it would pass the destructive challenge test. However, after passing the integrity test, the filter element can be placed in service and will provide the user with the results claimed by the filter manufacturer.

4.4.3 Beta (β) Rating System

While absolute ratings are clearly more useful than nominal ratings, a more recent system for expressing filtration rating is the assignment of Beta ratio values. Beta ratios are determined using the Oklahoma State University "OSU F-2 Filter Performance test." The test, originally developed for use on hydraulic and lubricating oil filters, has been adapted by many filter manufacturers for rapid semi-automated testing of filters for service with aqueous liquids, oils, or other fluids.

The Beta rating system is simple in concept and can be used to measure and predict the performance of a wide variety of filter cartridges under specified conditions. The rating system is based on measuring the total particle counts at several different particle sizes, in both the influent and effluent streams, a profile of removal efficiency emerges for any given filter.

The Beta value is defined as follows:

$$\beta = \frac{\text{number of particles of a given size and larger in influent}}{\text{number of particles of a given size and larger in effluent}}$$

where X is the particle size, in μm.

The percent removal efficiency at a given particle size can be obtained directly from the Beta value and can be calculated as follows:

$$\% \text{ removal efficiency} = \frac{\beta - 1}{\beta}(100).$$

The relationship between Beta values and percent removal efficiency is illustrated in Table 4.1.

Usually a $\beta = 5000$ can be used as an operational definition of an absolute rating.

The Beta values allow comparison of removal efficiencies at different particle sizes for different cartridges in a meaningful manner.

The type of filter medium, its rating (nominal, absolute, or Beta), and its solids loading are thus all important in selecting a filter. The

TABLE 4.1
Beta ratio and removal efficiency comparison

Filter	No. of Particle Per ml $\geq 10\,\mu m$		Beta Ratio B_{10}	Removal Efficiency %
	Influence	Effluent		
A	10,000	5000	2	50
B	10,000	100	100	99
C	10,000	10	1000	99.9
D	10,000	2	5000	99.98
E	10,000	1	10,000	99.99

designer should pay particular attention to the precise meaning of the manufacturer's rating.

4.5 Choosing the Proper Filter

Among the more important factors that must be taken into consideration when choosing a filter for a particular application are the size, shape, and hardness of the particles to be removed, the quantity of those particles, the nature and volume of the fluid to be filtered, the rate at which the fluid flows, whether the flow is steady, variable, and/or intermittent, the system pressure and whether that pressure is steady or variable, the available differential pressure, the compatibility of the medium with the fluid, the fluid temperature, the properties of the fluid, the space available for particle collection, and the degree of filtration required. Let's examine how some of these factors affect filter selection.

4.5.1 Nature of Fluid

The materials from which the medium, the cartridge hardware, and the housing are constructed must be compatible with the fluid being filtered. Fluids can corrode the metal core of a filter cartridge or a pressure vessel, and the corrosion will in turn contaminate the fluid being filtered. Thus, it is essential to determine whether a fluid is acid, alkali, aqueous, oil- or solvent-based, etc.

4.5.2 Flow Rate

Flow rate (the units of measurement are given as volume per unit of time, e.g., ml/min, gal/min) is dependent on two general parameters, pressure (P) and resistance (R). Flow rate depends directly on pressure and inversely on resistance. Thus, for a constant R, the greater the pressure, the greater the flow. For a constant P, lowering the resistance increases the flow.

Pressure can come from any number of sources and is usually expressed as pounds per square inch (psi) or bar gauge (barg). All other factors being equal, if the pressure on a fluid is increased, then the flow rate of that fluid will increase.

Viscosity is the resistance of a fluid to the motion of its molecules among themselves; in other words, a measure of the thickness or "flowability" of a fluid. Water, ether, and alcohol have low viscosities; heavy oils and syrup have high viscosities. Viscosity affects resistance directly. If all other conditions remain constant, doubling the viscosity in a filter system gives twice the original resistance to flow. Consequently, as viscosity increases, the pressure required to maintain the same flow rate increases. Centipoise (cp) is the unit of measurement comparing the viscosity of a fluid with that of water, which has a viscosity of 1 cp at 70 °F.

4.5.3 Temperature

The temperature at which filtration will occur can affect the viscosity of the fluid, the corrosion rate of the housing, and the filter medium compatibility. Viscous fluids generally become less viscous as temperature increases. If a fluid is too viscous, it may be advisable to preheat the fluid and to install heater bands in the filter. Thus, it is important to determine the viscosity of a fluid at the temperature at which filtration will occur.

High temperature also tends to accelerate corrosion and to weaken the gaskets and seals of filter housings. Very often disposable filter media cannot withstand high temperatures, particularly for prolonged periods of time. It is for this reason that one must often choose porous metal, cleanable filters.

4.5.4 Pressure Drop

Everything a fluid flows through or by contributes resistance to the flow of that fluid in an additive fashion. The pressure losses due to flow on the fluid through the tubing, piping, etc., couple with the pressure loss through the filter to produce resistance.

Resistance to flow through a clean filter will be caused by the filter housing, cartridge hardware, and filter medium. For a fluid of given viscosity, the smaller the diameter of the pores or passages in the medium, the greater the resistance to flow will be. When a fluid meets resistance in the form of a filter, the result is a drop in pressure downstream of that filter, and the measurement of the pressure drop across the filter is called the differential pressure, or ΔP. Thus, for all practical purposes the terms "pressure drop," "differential pressure," and "ΔP" are synonymous.

The more resistance a filter medium offers to fluid flow, the greater the differential pressure at constant flow will be. Since flow is always in the direction of the lower pressure, the differential pressure will cause fluid to flow. Thus, it is differential pressure that moves the fluid through the filter assembly and overcomes resistance to flow and ΔP.

In choosing a filter, one must provide a sufficient pressure source not only to overcome the resistance of the filter, but also to permit flow to continue at an acceptable rate as the medium plugs so as to use fully the effective dirt holding capacity of the filter. If the ratio of initial clean pressure drop through the filter to total available pressure is disproportionately high, unacceptable flow will quickly result even though the medium's capacity for collecting dirt has not been exhausted. When this occurs, the proper solution is usually to increase pump capacity or gravity head or, as an alternative, to reduce clean pressure drop by increasing filter size.

The maximum allowable cartridge pressure drop is the limit beyond which the filter might fail structurally should additional system pressure be applied to maintain adequate flow. This limit is always specified by the filter manufacturer.

In choosing a pressure source, one must take into consideration the resistance to flow of the filter—both constant resistance components (filter housing and element hardware) and the variable resistance components of the filter cake and medium. As filtration proceeds at constant flow, there will be an increase in pressure drop made up of a constant component and an increasing variable component. Eventually, the increasing pressure drop component becomes so large that either the filter clogs, which stops flow, or the filter is structurally damaged. Enough pressure drop should be available to satisfy both components at least to filter clogging.

If a pressure head exists downstream, as for example in an elevated receiver, this must be overcome without limiting the available pressure drop for the filter. In such cases, a check valve should be installed downstream of the filter to prevent reverse pressure from damaging the cartridge.

The pressure drop across the filter assembly can be reduced by increasing the size of the assembly. This allows an increased number of filter elements to be installed which in turn increases the total throughput. This is usually an economical approach for continuous processes where the increase in the larger filter assembly, and thus total throughput, is offset by the cost of using multiple smaller assemblies with the same throughput.

4.5.5 Surface Area

The life of most screen and fixed-pore structure filters is greatly increased as their surface areas are increased. To understand why this is so, let us look at two filters of identical medium (thus subject to the same pressure drop limit) that pass the same fluid at the same flow rate.

The first filter has a surface area of 5 ft^2 and collects a 0.005-in.-thick (128 µm) filter cake in a 24-hour period. After 24 hours most of the pores are plugged, the pressure drop is 75 psi, and the useful life of the filter has been exhausted.

Let's increase the surface area of the filter to 30 ft^2 and calculate the useful life. If a filter with a surface of 5 ft^2 collects a filter cake of

0.005 in 24 h, then at the same flow rate a filter of 30 ft^2 will collect that same filter cake in x h. Thus,

$$\frac{5}{24} = \frac{30}{x},$$
$$5x = (30)(24),$$
$$x = 144 \text{ h.}$$

While the 30-ft^2 filter has collected a filter cake of 0.005 in. in 144 h, its useful life will not be exhausted because the pressure drop will not have reached 75 psi (there are 6 times as many pores to plug: 30/5 = 6). Since the flow rate per ft^2 of filter area is in the ratio of 5/30, the pressure drop across the 30-ft^2 filter will be (5/30)(75 psi) = 12.5 psi. If the 30-ft^2 filter has a pressure drop of 12.5 psi in 144 h, then it will have a pressure drop of 75 psi in x h. Thus,

$$\frac{12.5}{144} = \frac{75}{x},$$
$$12.5x = (75)(144),$$
$$x = 864 \text{ h.}$$

The life of the 30-ft^2 filter is therefore 36 times that of the 5-ft^2 filter (864/24). If one calculates the square of the area ratio (30/5)^2, the answer is 36.

The benefit of opting for a filter assembly with a large surface area can be expressed as follows:

Let T = throughput (gallons) for a filter with area A (ft^2).

Then

$$T_1 = T_2 \left(\frac{A_1}{A_2}\right)^n,$$

where n is a life extension factor between 1 and 2. The life extension factor, n, will approach 2 provided that

1. The filter cake is not compressible. If the filter cake is compressible, n will tend to be nearer to 1.
2. The collected cake does not become a finer filter than the medium itself (i.e., collect finer solids as it builds up). To the extent that the filter cake acts as a finer filter than the medium itself, n will tend to approach 1.
3. The solids collected are relatively uniform in particle diameter.

From the above it is apparent that an increase in surface area will yield at least a proportional increase in service life. Under favorable circumstances, the ratio of service life may approach the square of the area ratio. In most cases, a filter user will save money in the long run by paying the higher initial cost of a larger filter assembly.

4.5.6 Void Volume

Void volume is always of great importance. All other factors being equal, the medium with the greatest void volume is most desirable because it will yield the longest life and lowest initial clean pressure drop per unit thickness. As the fiber diameter decreases, the void volume increases, assuming constant pore size. Other factors, however, such as strength, compressibility as pressure is applied (which reduces void volume), compatibility of the medium with the fluid being filtered, cost of medium, cost of constructing that medium into a usable filter, etc., must all be considered when designing a filter for a particular application.

4.5.7 Degree of Filtration

The filter chosen for a given application must be able to remove contamination from the fluid stream to the degree required by the process involved. Once the size of the contaminants to be removed has been determined, it is possible to choose a filter with the particle removal characteristics needed to do the job. Choosing a filter with a pore size finer than required can be a costly mistake. Remember, the finer the filtration, the more rapid the clogging and the higher the cost will be.

The filter selected must be able to retain particles removed from the subject fluid. Depth-type filters of the type whose pores can increase in size as pressure is increased are subject to unloading. With surface filters or fixed structure filters, one selects a medium that will not change its structure under system-produced stress.

4.5.8 Prefiltration

The purpose of a prefilter is to reduce overall operating cost by extending the life of the final filter. Extending final filter life may not in itself be sufficient to justify prefiltration; overall cost reduction is usually the principal consideration.

Field experience indicates that for most applications it is better to increase the final filter area rather than provide a prefilter. This is because increasing the final filter area always yields a longer cycle and lower operating costs. Doubling the area of the final filter will result in two to four times the life. On the other hand, installing settling devices, hydrocyclone desanders, or large pore space sand filters upstream of filters designed to remove very small particles is often a more economical solution than just increasing the size of the final "polishing" filter time.

4.5.9 Coagulants and Flocculation

Suspended matter in water may contain very small particles that will not settle out by gravity or that may pass through filters. These particles may be removed by a coagulation-and-flocculation process. Coagulation is the process of destabilization by charge neutralization, and flocculation is the process of bringing together the destabilized or coagulated particles to form a larger agglomerate, or floc.

Coagulation and flocculation results are difficult to predict based on a water analysis; therefore, laboratory jar tests are performed to simulate the coagulation and flocculation condition. The laboratory data are then used to determine the basis for design and efficient operation. The tests are run to establish:

- Optimum pH for coagulation,
- Most effective coagulation and coagulation aid,
- Most effective coagulation dosage and order of chemical addition,
- Coagulation and flocculation time,
- Settling time or flocculation time,

Chemicals used include:

- Chlorine,
- Bentonite (for low-turbidity water),
- Primary inorganic coagulants,
- pH adjustment chemicals, and
- Polyelectrolytes.

Chlorine addition may assist coagulation by oxidizing organic contaminants that have dispersing properties. Waters with high organic content require high coagulant demand. Chlorination prior to addition of coagulant feed may reduce the required coagulant dosage.

The term "polyelectrolytes" refers to all water-soluble organic polymers used for clarification of water by coagulation. The available water-soluble polymers may be classified as anionic, cationic, or approaching neutral charge. They are typically long-chain, high-molecular-weight polymers with many charge sites to aid in coagulation and flocculation. Violent mixing of polyelectrolytes may break the chains and cause them to be less effective. However, some mixing is required to ensure that the chemical and solids come into contact. Turbulent flow in piping provides sufficient mixing if the chemical is injected far enough upstream of the equipment.

Chemicals can be added to the water in clarifiers, which are tanks that contain mixers that cause sufficient turbulence to create contact between the chemical and the solids. Coagulation/flocculation chemicals can also be added in flotation units to aid in attaching gas bubbles to the solid particles. Polyelectrolytes added to the feed stream to filtration units have proven effective at increasing filtration efficiency.

4.6 Measuring Water Compatibility

Scale deposits are usually salts or oxides of calcium, magnesium, iron, copper, and aluminum. Common scale deposits may consist of calcium carbonate ($CaCO_3$), calcium phosphate ($CaPO_4$), calcium silicate ($Ca_2Si_2O_4$), calcium sulfate ($CaSO_4$), or magnesium hydroxide [$Mg(OH)_2$], magnesium phosphate ($MgPO_4$), and magnesium silicate ($Mg_2Si_2O_4$).

The tendency of water to form scale or cause corrosion is measured by either the Langelier Scaling Index (LSI), which is also called the Saturation Index, or the Ryznar Stability Index (RSI), which is also called the Stability Index (Table 4.2).

TABLE 4.2
Saturation index

To determine:	
pCa	Locate ppm value for Ca as $CaCO_3$ on the ppm scale. Proceed horizontally to the left diagonal line down to the pCa scale.
pAlk	Locate ppm value for "M" Alk as $CaCO_3$ on the ppm scale. Proceed horizontally to the right diagonal line down to the pAlk scale.
Total solids	Locate ppm value for total solids on the ppm scale. Proceed horizontally to the proper temperature line and up to the "C" scale.
Example:	
Temp. = 140 °F, pH = 7.80	pCa = 2.70
Ca hardness = 200 ppmw	pAlk = 2.50
M alkalinity = 160 ppmw	C at 140 °F = 1.56
Total solids = 400 ppmw	Sum = pH 3 = 6.76
	Actual pH = 7.80
	Difference = 1.04
	= Saturation index

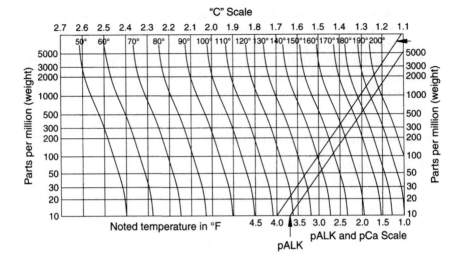

FIGURE 4.4. Langelier Saturation Index Chart (reprinted from GPSA Engineering Data Book, courtesy of Betz Laboratories, Inc.).

The LSI deals with the conditions at which given water is in equilibrium with calcium carbonate and provides a qualitative indication of the tendency of calcium carbonate to deposit or dissolve. The index is determined by subtracting from the actual pH of the water sample (pHA) a computed value (pHS) based on the ppm of calcium hardness as $CaCO_3$, alkalinity hardness as $CaCO_3$, and total solids, as shown in Figure 4.4. If the index is positive, calcium carbonate will tend to deposit. If it is negative, calcium carbonate will tend to dissolve.

4.6.1 Saturation Index

The Stability Index (RSI) is given by

$$RSI = 2pHS - pHA.$$

where:

pHA = actual pH of water sample
pHS = computed value from Figure 4.4 as explained above

When the index is less than 6, scaling can be expected; an index between 6 and 7 indicates a stable water. A pH greater than 7 indicates potential corrosion problems.

Scaling may be controlled by the following: blow-down to limit build-up of solids concentrations; acid treatment to reduce the water alkalinity; and use of commercial scale inhibitors such as polyphosphates, phosphonates, and polymers.

4.7 Solids Removal Equipment Description

This section describes several different processes designed to remove suspended solids or dissolved gases from water. As stated, the most common reason for such water treatment is associated with water injection; therefore, the subject of water injection will also be discussed throughout this chapter. The principles and equipment involved can also be applied when the water must be treated for other reasons.

Figure 4.5 shows a schematic of steps that may be required to prepare water for injection. As shown in the figure, the choice of process is affected by the water source.

Almost every oil or gas production facility must deal with some produced water at some time during its production life. In many facilities the water may be disposed of once its hydrocarbon content is reduced to acceptable levels. When water is to be injected, it may be necessary to treat the water to attain low levels of oil (on the order of 25–50 mg/l) to prevent impairment of the injection formation or downstream equipment.

Typically, produced water may require filtration to remove dispersed oil. Filtering will remove all but a very small amount of the dispersed oil. It will be necessary, however, to clean the water to less than approximately 50 mg/l suspended oil so that oil does not create plugging problems in the filters. The small amount of suspended oil in the water will be removed by the filter and will contaminate the filter backwash.

Surface water is a common water source for water-flood or other injection projects. If surface water is available, it may be cheaper to obtain than subsurface water. However, surface water may require more treatment than other water sources, and, since it is fresh, it may cause swelling of clays in some formations.

The surface water should be free of large contaminants such as plant or marine life. The strainer in Figure 4.5 is intended to prevent such material from entering treatment facilities.

Surface water is exposed to the atmosphere; therefore, it contains dissolved oxygen. Typically, the oxygen requires removal to minimize corrosion and bacterial growth. The oxygen concentration within the water will vary, depending on the water temperature; therefore, de-aeration equipment must be designed to remove the maximum anticipated oxygen content. Oxygen concentrations of approximately 8 ppm are typical for most surface waters.

Chemical injection of biocides is of particular importance for treating surface water. The water contains microscopic marine life, bacteria, plankton, and algae, which cannot be allowed to continue to grow within treating equipment.

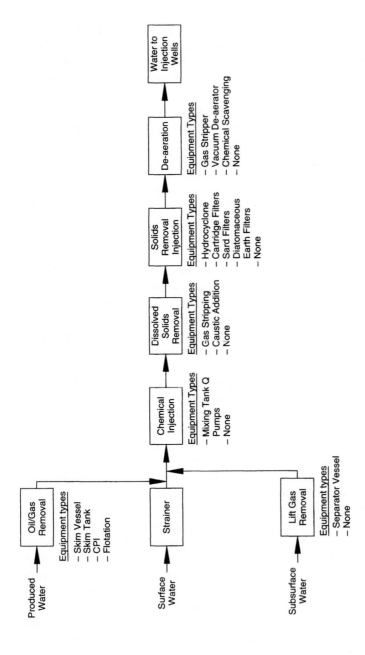

FIGURE 4.5.   Water injection system treatment steps and equipment types.

Injection of other chemicals may also be required for other reasons. Corrosion inhibitor may be required to minimize deterioration of the surface facilities. Bactericide may be used on either a periodic or a continuous basis to minimize bacteria growth. Chemicals such as oxygen scavengers may be injected to react with contaminants to remove or change them. Similarly, chemicals may be injected to prevent the formation of scale within the surface equipment or the precipitation of solids under reservoir conditions.

Subsurface water from source water wells may also be used. Typically, if water is available at all, the water zone will be close to the surface, compared with hydrocarbon zones. Therefore, the cost of drilling and maintaining source water wells is typically much lower than the same cost for producing wells. However, the water rarely will flow to the surface under pressure; therefore, it must be pumped or gas lifted.

If the water is gas lifted to the surface, separation facilities will be required. Typically, a simple two-phase separator is sufficient, since small levels of dissolved natural gas in the water are not harmful to the equipment or the injection formation. If the gas used to lift the water contains acid gases or oxygen, then some treating may be necessary to remove or neutralize these harmful gas components.

Subsurface water is typically the least expensive source water to treat. However, the cost of drilling source water wells and the pumping or gas lifting expenses may make this the most expensive water to obtain.

Water from any source contains dissolved minerals and salts in addition to the suspended solids. Generally, these dissolved salts will remain in solution and are thus not a problem. However, if produced water is to be mixed with other source water as part of a waterflood, the water compatibility should be verified. Mixing other waters with produced water or changing the produced water's pH or temperature may cause dissolved solids to precipitate.

The compatibility of the injected water and the water in the injection formation should also be checked to ensure that the conditions within the reservoir will not cause scale to form. Chemicals may be needed to inhibit scale formation under down-hole conditions.

Filtration to remove suspended solids, no matter what the source, is intended to minimize plugging of the formation. Solids that are in the injected water and are larger than a certain size may plug the formation at the well bore, causing the surface injection pressure to rise or the flow rate injected to fall.

The degree of filtration required depends on the permeability and pore size of the injection formation. The final selected filtration design should be one that will minimize formation plugging and reduce the frequency of remedial well work. This selection is an economic decision balancing the cost of remedial well work against the cost of the filter system. In cases where the water is being injected into disposal wells, filtration requirements might be relaxed. Disposal wells are typically less expensive to drill or work over and more readily fractured than wells injecting the water into a producing formation. Therefore, the economic risks of plugging a disposal well may be less than those associated with a well injecting the water into the producing formation.

The test to check for compatibility of injection water with the receiving formation may consist of

- a chemical analysis of the proposal injection water to indicate basic cations and anions present,
- a core plug test to determine the maximum particle size that can be injected into the formation without undue plugging, and
- a core plug test to determine and establish permissible injection rates and pressures.

4.7.1 Gravity Settling Tanks

The simplest treating equipment for removing solids from water is a gravity settling tank or vessel, which may be designed in either a vertical or horizontal configuration. In vertical settling tanks, the solid particles must fall countercurrent to the upward flow of the water. A typical vertical gravity settling vessel is shown in Figure 4.6. The water enters the vessel and flows upward to the water outlet. Solids fall countercurrent to the water and collect in the bottom. As shown, large-diameter vessels or tanks should have spreaders and collectors to distribute the water flow and minimize short-circuiting.

For small-diameter gravity settling vessels, the collected solids may be removed by periodically opening the sand drain shown in Figure 4.7. The use of a cone bottom rather than an elliptical head allows more complete removal of solids through the drain. Typically, an angle of 45°–60° from the horizontal is used for the cone to overcome the angle of internal resistance of the sand and allow natural flow of solids when the drain is opened.

Any flash gases that evolve from the water leave the settling vessel through the gas outlet at the top of the vessel. The volume of flash gas must be kept to a minimum so the gas does not adversely affect the removal of small solids particles. If large amounts of gas are flashed, the small gas bubbles can adhere to solids particles and carry them to the water surface. The solids then may be carried out the water outlet.

In horizontal settlers, the solids fall perpendicular to the flow of the water, as shown in Figure 4.7. The inlet is often introduced above

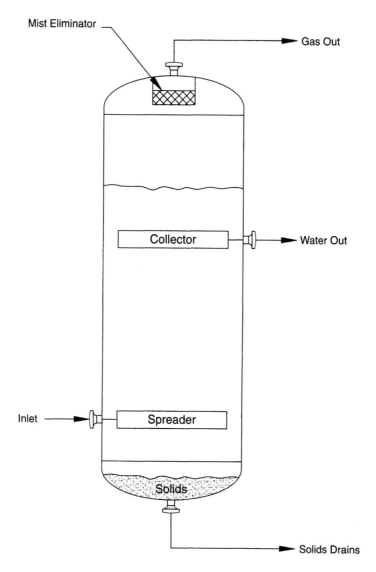

FIGURE 4.6. Schematic of vertical gravity settling vessel.

the water section so that flash gases may be separated from the water prior to separating the solids from the water.

The collected solids must be periodically removed from the vessel; thus, several drains may be placed along the length of a horizontal vessel. Since the solids will have an angle of repose of 45°–60°, the drains must be spaced at very close intervals and operated frequently to prevent plugging. The addition of sand jets in the vicinity of each drain to fluidize the solids while the drains are in operation is

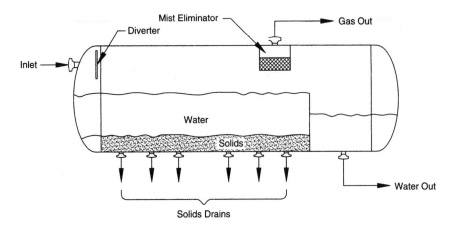

FIGURE 4.7. Schematic of horizontal gravity settling vessel.

expensive, but sand jets proved successful in keeping drains open. Alternatively, the vessel may have to be shut down so that solids may be manually removed through a manway. Although effective, this method can be extremely expensive and time-consuming.

Horizontal vessels are more efficient at solids separation because the solid particles do not have to fall countercurrent to the water flow. However, other considerations, such as the difficulty of removing solids, must be kept in mind when such a configuration is selected. Horizontal vessels require more plan area to perform the same separation as vertical vessels. Most small horizontal vessels have less liquid surge capacity. For a given change in liquid surface elevation, there is typically a larger increase in liquid volume for a horizontal vessel than for a vertical vessel sized for the same flow rate. However, the geometry of most small horizontal vessels causes any high-level shutdown device to be located close to the normal operating level. In large-diameter [greater than 1.8 m (6 ft)] horizontal vessels and in vertical vessels, the shutdown could be placed much higher, allowing the level controller and dump valve more time to react to a surge.

It should be pointed out that vertical vessels have some drawbacks that are not process-related; these must be considered in making a selection. For example, the relief valve and some of the controls may be difficult to service without special ladders and platforms. The vessel may have to be removed from a skid for trucking due to height restrictions.

The choice of a pressure vessel versus an atmospheric tank for a solids-settler depends on the overall needs of the system. Although pressure vessels are more expensive than tanks, they should be considered when potential gas blow-by through an upstream vessel dump system could create too much back pressure in an atmospheric tank's vent system; or when the water must be dumped to a higher elevation for further treating and a pump would be needed if an atmospheric tank were installed.

For gravity settling of solids, water retention time does not directly affect the solids removal, and only settling theory must be considered. Some small retention time is required for evolved gases to flash out of solution and reach equilibrium. This process usually requires less than 30 sec; therefore, retention time criteria rarely govern vessel size.

Gravity settling is useful for settling large-diameter solids. It is normally used where there is a high solids (greater than 50 μm) flow rate of large-diameter solids that might otherwise quickly overload equipment designed to separate smaller-diameter solids from the liquid stream.

4.7.2 Horizontal Cylindrical Gravity Settlers

The required diameter and length of a horizontal cylindrical settler can be determined from Stokes' law as follows:

Field units

$$dL_{eff} = 1000 \frac{\beta_w Q_w \mu_w}{\alpha_w (\Delta SG) d_m^2}, \tag{4.4a}$$

SI units

$$dL_{eff} = 1.2 \times 10^9 \frac{\beta_w Q_w \mu_w}{\alpha_w (\Delta SG) d_m^2}, \tag{4.4b}$$

where

d = vessel's internal diameter, ft (m),
L_{eff} = effective length in which separation occurs, ft (m),
Q_w = water flow rate, BWPD (m³/h),
μ_w = water viscosity, cp,
d_m = particle diameter, μm,
ΔSG = difference in specific gravity between the particle and water relative to water,
β_w = fractional water height within the vessel (h_w/d),
α_w = fractional cross-sectional area of water,
h_w = water height, in. (m).

Equations (4.4a) and (4.4b) assume a turbulence and short-circuiting factor of 1.8. Any combination of d and L_{eff} that satisfies this equation will be sufficient to allow all particles of diameter d_m or larger to settle out of the water.

The fractional water height and fractional water cross-sectional areas are related by the following equation:

$$\alpha_w = (1/180)\cos^{-1}[1 - 2\beta_w] - (1/\pi)[1 - 2\beta_w]$$
$$\sin[\cos^{-1}(1 - 2\beta_w)]. \tag{4.5}$$

By selecting a fractional water height within the vessel, one may calculate the associated fractional cross-sectional area using Equation (4.5); the resulting values may then be used in Equations (4.4a) and (4.4b).

In addition to the settling criteria, a minimum retention time should be provided to allow the water and flash gases to reach equilibrium. Typically, retention times to reach equilibrium are less than 30 sec. Although retention time requirements rarely govern the size of a settling vessel, these requirements should be checked. To ensure that the appropriate retention time has been provided, the following equation must also be satisfied when selecting d and L_{eff}:

Field units

$$d^2 L_{eff} = \frac{(t_r)_w Q_w}{1.4\alpha_w}. \tag{4.6a}$$

SI units

$$d^2 L_{eff} = 21,000\frac{(t_r)_w Q_w}{1.4\alpha_w}. \tag{4.6b}$$

The choice of the correct diameter and length can be obtained by selecting various values for d and L_{eff} for Equations (4.4a)–(4.6b). For each d the larger L_{eff} must be used to satisfy both equations.

The relationship between the L_{eff} and the seam-to-seam length of a settler is dependent on the settling vessel's internal physical design. Some approximations of the seam-to-seam length may be made based on experience as follows:

$$L_{ss} = (4/3)L_{eff}, \tag{4.7}$$

where L_{ss} is the seam-to-seam length. This approximation must be limited in some cases, such as vessels with large diameters. Therefore, the L_{ss} should be calculated using Equation (4.7), but it must exceed the value calculated using the following equations:

Field units

$$L_{ss} = L_{eff} + 2.5, \tag{4.8a}$$

where $L_{eff} < 7.5$ ft.

SI units

$$L_{ss} = L_{eff} + 0.76. \tag{4.8b}$$

Field units

$$L_{ss} = L_{eff} + \frac{d}{24}.$$ (4.9a)

SI units

$$L_{ss} = L_{eff} + \frac{d}{2000}.$$ (4.9b)

Equations (4.8a) and (4.8b) only govern when the calculated L_{eff} is less than 2.3 m (7.5 ft). The justification of this limit is that some minimum vessel length is always required for smoothing the water inlet and outlet flow. Equations (4.9a) and (4.9b) governs when one half the diameter in feet exceeds one third of the calculated L_{eff}. This constraint ensures that even flow distribution can be achieved in short vessels with large diameters.

4.7.3 Horizontal, Rectangular Cross-Sectional Gravity Settlers

Similarly, the required width and length of a horizontal tank of rectangular cross section can be determined from Stokes' law as

Field units

$$WL_{eff} = 70 \frac{Q_w \mu_w}{(\Delta SG)d_m^2},$$ (4.10a)

SI units

$$WL_{eff} = 9.7 \times 10^5 \frac{Q_w \mu_w}{(\Delta SG)d_m^2},$$ (4.10b)

where

W = width, m (ft),
L_{eff} = effective light in which separation occurs, ft (m).

Equations (4.10a) and (4.10b) assume a turbulence and short-circuiting factor of 1.9. Note that Equations (4.10a) and (4.10b) are independent of height because the particle settling time and the water retention time are both proportional to the height. Typically, the height is limited to one half of the width to promote even flow distribution.

As before, an equation may be developed to ensure that sufficient retention time is provided. If the height-to-width ratio is set, then the following retention time equation applies:

Field units

$$W^2 L_{eff} = \frac{0.004(t_r)_w Q_w}{\gamma},$$ (4.11a)

SI units

$$W^2 L_{\text{eff}} = \frac{(t_r)_w Q_w}{60\gamma},$$ (4.11b)

where

γ = height-to-width ratio (H_w/W),
H_w = height of the water, ft (m).

As with horizontal cylindrical settlers, the relationship between L_{eff} and L_{ss} depends on the internal design. Three approximations of the L_{ss} of rectangular settling vessels may be made using Equations (4.4a) and (4.4b). However, the L_{ss} must be limited by Equation (4.5) and the following:

$$L_{\text{ss}} = L_{\text{eff}} + \frac{W}{2}.$$ (4.12)

As before, the L_{ss} should be the largest of Equations (4.7)–(4.8b) and (4.12).

4.7.4 Vertical Cylindrical Gravity Settlers

The required diameter of a vertical cylindrical tank can be determined by setting the settling velocity equal to the average water velocity as follows:

Field units

$$d^2 = 6700 \, F \frac{Q_w \mu_w}{(\Delta\text{SG})d_m^2},$$ (4.13a)

SI units

$$d^2 = 6.5 \times 10^{11} \, F \frac{Q_w \mu_w}{(\Delta\text{SG})d_m^2},$$ (4.13b)

where F is the factor that accounts for turbulence and short-circuiting. For small-diameter settlers, the short-circuiting factor should be equal to 1.0. Settlers with diameters greater than 48 in. (1.22 m) require a larger value for F. Inlet and outlet spreaders and baffles affect the flow distribution in large settlers and therefore affect the value of F. It is recommended that, for large-diameter settlers, F should be set equal to $d/48$. Substituting this into Equations (4.13a) and (4.13b) gives the following:

Field units

$$d = 140 \frac{Q_w \mu_w}{(\Delta\text{SG})d_m^2},$$ (4.14a)

SI units

$$d = 5.3 \times 10^9 \frac{Q_w \mu_w}{(\Delta SG) d_m^2},$$
(4.14b)

where $d > 48$ in. (1.22 m). Equations (4.14a) and (4.14b) apply only if the settler diameter is greater than 48 in (1.22 m). For smaller settlers, Equations (4.13a) and (4.13b) should be used and F should equal 1.0.

The height of water column in feet can be determined for a selected d from retention time requirements:

Field units

$$H = 0.7 \frac{(t_r)_w Q_w}{d^2}$$
(4.15a)

SI units

$$H_w = 21,000 \frac{(t_r)_w Q_w}{d^2},$$
(4.15b)

where H is the height of water, in ft (m).

4.7.5 Plate Coalescers

The equations for sizing the various configurations are identical to those presented in Chapter 3 and can be used directly, where d_m is the diameter of the solid particle (and not the oil droplet diameter) and ΔSG is the difference in specific gravity between the solid and water (and not between oil and water).

Plate coalescers are not addressed in this section because they have a tendency to plug and are thus not recommended for solids separation.

4.7.6 Hydrocyclones

Hydrocyclones, also called desanders or desilters, operate by directing the water into a cone through a tangential inlet that imparts rotational movement to the water. Figures 4.8A and 4.8B show a hydrocyclone cone and an assembly of eight cones.

The rotary motion generates centrifugal forces toward the outside of the cone, driving the heavy solids to the outer perimeter of the cone. Once the particles are near the wall, gravity draws them downward to be rejected at the apex of the cone. The resulting heavy slurry is then removed as "underflow." The clear water near the center of the vortical motion is removed through an insert at the centerline of the hydrocyclone, called a "vortex finder," and passes out as "overflow" through the top of the cone.

The advantage of hydrocyclones is that the centrifugal forces separate particles without the need for large settling tanks. Operationally, hydrocyclones are good at removing solids with diameters of approximately 35 μm and larger.

A major drawback of hydrocyclones is that during upsets in flow or pressure drop, the rotary motion in the cone may be interrupted, possibly causing solids to carry over into the overflow liquid. Other drawbacks are wear problems, large pressure drops, and limited ability

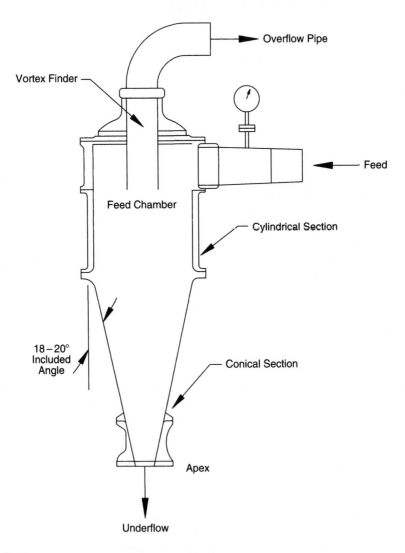

FIGURE 4.8. (A) Schematic of a hydrocyclone.

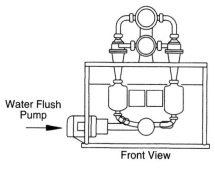

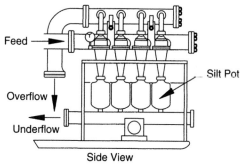

FIGURE 4.8. (B) Schematic of a hydrocyclone.

to handle surges in flow. Some manufacturers offer replaceable liners to handle wear problems.

Hydrocyclones are rarely used as the only solids removal device, although they can remove very high loadings of solids, making them very useful as a first step in solids removal. If filters are used as a second step, the hydrocyclone can greatly lengthen the filters' cycle time. At the same time, the filters can provide removal of the smaller-diameter solids and protect against carryover from the hydrocyclones during upsets.

The ability of a hydrocyclone to separate a certain diameter solid particle (fineness of separation) is affected by the differential pressure between the inlet and overflow, the density difference between the solid particles and the liquid, and the geometry and size of the cone and inlet nozzle. The pressure drop through the cone is the critical variable in terms of affecting fineness of separation and is itself a function of flow rate. Thus, the lower the flow rate, the lower the pressure drop and the coarser the separation. A minimum flow must be maintained through each cone to create the required pressure drop and rotary motion to ensure separation.

Typically, hydrocyclones are operated with a 25–50 psi (140–275 kPa) pressure drop.

Many theoretical and empirical equations have been proposed for calculating fineness of separation. All reduce to the following form for a hydrocyclone of fixed proportions:

$$d_{50} = K \left[\frac{D^3 \mu}{Q_w (\Delta SG)} \right]^{\frac{1}{2}},$$
(4.16)

where

D = major diameter of hydrocylone cone, ft(m)
d_{50} = solid particle diameter that is recovered 50% to the overflow and 50% to the underflow, μm,
μ = slurry viscosity, cp (Pas),
Q_w = slurry flow rate, BPD (m³/h),
ΔSG = difference in specific gravity between the solid and the liquid,
K = proportionally and shape constant.

The diameter of the solid particle that is recovered 1–3% to the overflow and 97–99% to the underflow is

$$d_{99} = 2.2 d_{50}.$$
(4.17)

The flow rate through a hydrocyclone of fixed proportions handling a specified slurry is given by

$$Q_w = K^I \Delta P \frac{1}{2},$$
(4.18)

where

Q_w = flow rate, BWPD (m³/h),
K^I = proportionally and shape constant,
ΔP = pressure drop, psi (kPa).

Equations (4.16)–(4.18) can be used to approximate the performance of a hydrocyclone for different flow conditions, if its performance is known for a specific set of flow conditions.

Solids discharge in the underflow slurry is performed in either an open or a closed system. With the open system, the slurry is rejected through an adjustable orifice at the apex of the cone to an open trough. The orifice can be adjusted to regulate the flow rate of the water leaving with the solids. The open system can allow oxygen entry into the system.

In the closed system, a small vessel called a "silt pot" is connected to the apex, which remains open. A valve is located at the bottom of the silt pot and is normally closed. Solids pass through the apex and collect in the bottom of the silt pot. The valve at the bottom of the silt pot is opened periodically to reject the solids. The opening and closing of this valve can be manual or automatic.

Hydrocyclone units may be put online individually, thus providing some ability to account for changes in flow rate. When specifying a hydrocyclone unit, the design engineer must provide the following information:

- total water flow rate,
- particle size to be removed and the percentage of removal required,
- concentration, particle size distribution, and specific gravity of particles in the feed,
- design working pressure of the hydrocyclone, and
- minimum pressure drop available for the hydrocyclone.

With this information the designer can select equipment from various manufacturers' catalog descriptions.

4.7.7 Centrifuges

Centrifuges can be used to separate low-gravity solids or very high percentages of high-gravity solids. The principle involved is the same as in a hydrocyclone in that centrifugal force rapidly separates solids from the liquid. Centrifuges typically require extensive maintenance and can handle only small liquid flow rates. For these reasons centrifuges are not commonly used in water treating applications.

4.7.8 Flotation Units

It is possible to remove small particles using dispersed or dissolved gas flotation devices. These units are primarily used for removing suspended hydrocarbons from water. Gas is normally dispersed into the water or released from a solution in the water, forming bubbles approximately 30–120 μm in diameter. The bubbles form on the surfaces of the suspended particles, creating particles whose average density is less than that of water. These rise to the surface and are mechanically skimmed. In the feed stream, chemicals called "float aids" are normally added to the flotation unit to aid in coagulation of solids and attachment of gas bubbles to the solids. The optimum concentration and chemical formulation of float aids are normally determined from batch tests in small-scale plastic flotation models on-site. Because of the difficulty of predicting particle removal efficiency with this method, it is not normally used to remove solids from water in production facilities.

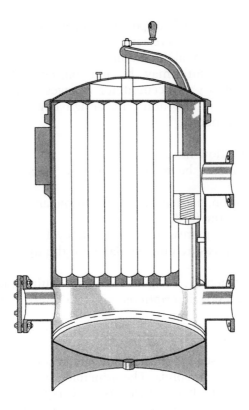

FIGURE 4.9. Cartridge filter vessel. (Courtesy of Perry Equipment Corp.)

4.7.9 Disposal Cartridge Filters

Cartridge filters are simple and relatively lightweight, and they can be used to meet a variety of filtration requirements. A typical cartridge filter vessel is shown in Figure 4.9. The water enters the top section and must flow through one of the filter elements to exit through the lower section of the vessel. The top head of the vessel is bolted so that the cartridges can be changed when the pressure drop across them reaches an upper limit. A relief valve can be included in the vessel to prevent excessive differential pressure between the upper and lower sections of the vessel.

Filter cartridges are available in a wide variety of materials, and they provide a range of performance options. Cartridges are available with manufacturers' particle size ratings from 0.25 μm to any larger particle size. When selecting a filter cartridge, the designer must determine what the manufacturer's rating actually means in terms of removal percentage.

Filter cartridge solids removal performances and allowable flow rates vary greatly from manufacturer to manufacturer, even if the cartridges are made of the same material. Therefore, it is difficult to develop generalized relationships between the water flow rate and filter area. As a result, it is necessary to rely on manufacturers' information when selecting and sizing a cartridge filter system.

In designing a water treatment system that includes cartridge filters, it may be desirable to select a fixed-pore filter medium and absolute rated filters. The fixed-pore cartridges provide more consistent particle removal efficiencies from one cartridge to the next than do nonfixed-pore cartridges. The fixed-pore type also prevents solids unloading and media migration during periods of high differential pressure. Fixed-pore filters are usually given absolute ratings by their manufacturers.

Nonfixed-pore cartridges may be used, but the differential pressure across the filters must be monitored closely. High differential pressures may cause solids unloading and media migration. If either occurs, the pressure drop through the filter will decrease and may be below the limit when the cartridge is scheduled to be changed. Therefore, the operator checking the pressure drop will believe that the cartridges are functioning correctly, even though large amounts of solids may have been released to the downstream water.

Solids unloading may be avoided by using a high differential pressure switch to continuously monitor the pressure drop or by changing the cartridges when the pressure drop is still small compared to the maximum pressure drop recommended by the manufacturer. The resulting frequent changing of the cartridges may result in excessive operating costs if the early change-out method is used.

Typically, cartridge filters have low solids-loading limits, so the cartridges can absorb only a relatively small amount of solids before they must be changed. Manufacturers have developed special cartridges to improve solids loading. Pleated construction of a thin filter medium such as paper or cotton fabric greatly increases the effective filter surface area of the cartridge. The increased surface area provides for higher flow rates and solids-loading capacities than a cylindrical cartridge of the same medium.

Some cartridges use a multi-layered design of media such as fiberglass, which provides in-depth filtration. The layers of media have progressively smaller pores as the water moves from the outside to the inside of the cartridge. As the pore size changes, particles are trapped at different depths within the filter, allowing higher solids loadings but typically decreasing flow rates slightly.

Since cartridge filters have low solids-loadings capacities, it is common to install primary solids removal equipment upstream of the cartridge filters. Typical systems include either a hydrocyclone or a sand filter followed by the cartridge filter. The upstream equipment removes the larger solids and reduces the amount of solids that the cartridges must remove, therefore extending the time between cartridge changes.

A spare filter vessel may be provided so that cartridges may be changed without reducing water flow rates. Any number of vessels can be used to provide the required number of cartridges, but the most common system arrangements include three 50% vessels or four 33% vessels. The number of filter vessels selected depends on a cost analysis and on operating preference.

Other factors to consider in the selection of cartridge filters are the type of filter medium and its characteristics. As an example, polypropylene cartridges are a better selection than cotton for water service, since cotton swells. The compatibility of filter membranes and binders with chemical additives or impurities in the water should be checked. The designer should contact specific manufacturers for detailed information.

When specifying a cartridge filter unit the following information should be included:

- maximum water flow rate,
- particle size to be removed by filtration and the percentage of removal required,
- solids concentration in the inlet water,
- design working pressure of the filter vessel, and
- maximum pressure drop available for filtration.

4.7.10 Backwashable Cartridge Filters

Backwashable cartridge filters are available in a variety of designs using metal screens, permeable ceramic, or consolidated sand as a filter medium. Filters of this type are simple and lightweight like the disposable cartridge filters, but they have the additional advantage of being backwashable. The media used in backwashable filters typically provide filtration of particles between 10 and 75 µm.

Backwashable cartridge filters have low solids-loading limits; therefore, they have potentially short intervals between backwash cycles. It is important not to expose backwashable filters to differential pressures over approximately 170 kPa because the particles may become too deeply imbedded in the pores to be removed by backwashing. With proper maintenance and repeated backwashing, this type of filter may last up to 2 years.

Regeneration or backwashing involves flowing clean water through the filter in the opposite direction of the normal filtration. Backwashable filters often require an acid backwash as well. The solids trapped in the filter media are then forced out of the filter and carried away with the backwash fluid. This process is quicker and may be less costly than changing cartridges. The flow rate of fluid required for backwash is specified by the manufacturer.

The disadvantage of this system is that filtered water must be stored and then pumped through the filter. The resulting backwash fluid must then be directed to another storage medium. A method and equipment for disposing of the backwash fluid, which can be contaminated with oil or acid used in the backwash cycle, must also be provided.

Filters of this type are available in a variety of designs, including the cartridge filter vessel in Figure 4.9. Alternatively, each cartridge may be in a separate housing and the housings may be manifolded on a skid. With the manifold design, it is possible to backwash individual filters while the other filters continue to operate normally.

The designer should contact manufacturers for detailed information on selecting filters of this type. When specifying a backwashable cartridge filter, the designer should include the following:

- maximum water flow rate,
- particle size to be removed by filtration and the percentage of removal required,
- solids concentration in the inlet water,
- design working pressure of the filter vessel, and
- maximum pressure drop available for filtration.

4.7.11 Granular Media Filters

The terms "granular media filter" and "sand filter" refer to a number of filter designs in which fluid passes through a bed of granular medium. Typically, these filters consist of a pressure vessel filled with the filter media, as shown in Figure 4.10. Media support screens prevent the media solids from leaving the filter vessel.

The water to be filtered may flow either downward (down-flow) or upward (up-flow) through the media. As the water passes through the media, the small solids are trapped in the small pores between the media particles. Down-flow filters may be designed as either "conventional" (see Figure 4.11) or "high flow rate" (see Figure 4.12). Conventional down-flow filters are normally designed for an approximate flow-rate range of 1–8 gpm/ft^2 (2.5–20 m^3/h m^2), while high-flow-rate types may have flow rates as high as 20 gpm/ft (249 m^3/h m^2). Up-flow filters (see Figure 4.13), on the other hand, are limited to less than 8 gpm/ft^2 (20 m^3/h m^2) because higher flow rates may fluidize the media bed and, in effect, backwash the media.

The advantage of the high-flow-rate filter over the conventional down-flow filter is that, at higher velocities, a deeper penetration of

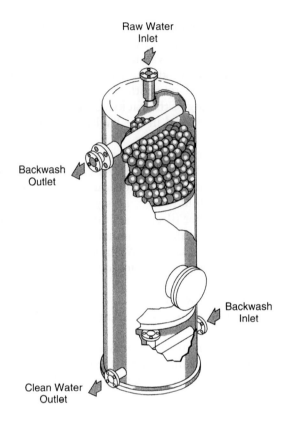

Raw Water
Inlet

Backwash
Outlet

Backwash
Inlet

Clean Water
Outlet

FIGURE 4.10. Down-flow granular media filter. (Courtesy of CE Natco.)

the bed is achieved, allowing a higher solids loading (weight of solids trapped per cubic foot of bed). This factor results in both a longer interval between backwashing and a smaller-diameter vessel. The disadvantage is that, with deeper penetration, inadequate backwashing may allow formation of permanent clumps of solids that gradually decrease the filter capacity. If fouling is severe, the filter media must be chemically cleaned or replaced.

Granular media filters must be cleaned periodically by backwashing to remove filter solids. The process involves fluidizing the bed to eliminate the small pore spaces in which solids were trapped during filtration. The small solids are then removed with the backwash fluid through a media screen that prevents loss of media solids. The filter media may be fluidized by flowing water upward through the filter at a high rate or by introducing the water through a nozzle that produces high velocities and turbulence within the filter vessel. Recycle pumps may be used to pump water through the fluidization nozzle to decrease the total water volume required to fluidize the

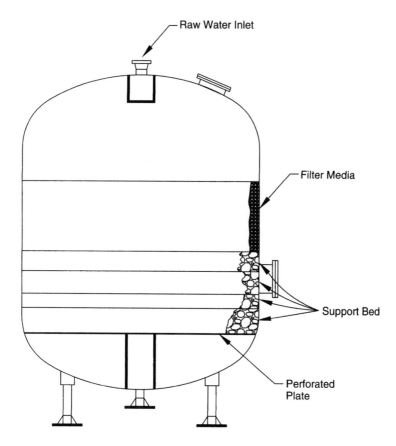

FIGURE 4.11. Conventional graded bed filter.

filter media. As with backwashable cartridge filters, the backwash fluid must be collected for disposal.

The backwashing process is usually initiated because of a high pressure drop through the filter. Alternatively, the filter may be backwashed on a regular schedule, provided the pressure drop limit is not exceeded between backwash cycles. The cycle time for a sand filter depends on the water's solids content and the allowable solids loading of the individual filter. Conventional down-flow filters with flow rates of less than 8 gpm/ft^2 (20 m^3/h m^2) typically can remove 1/2–1 1/2 lb/ft^2 (2.4–7.3 kg/m^2) of solids of filter media before backwashing. High-flow-rate filters may remove up to 4 lb/ft^2 (19.5 kg/m^2) prior to backwashing because the high water velocity forces small solids farther into the media bed, increasing the effective depth of the filter and thus the number of pores available to trap solids. Up-flow filters may remove up to 6 lb/ft^2 (29.3 kg/m^2) because the upward flow loosens and partially fluidizes the bed, allowing greater penetration by the small solids.

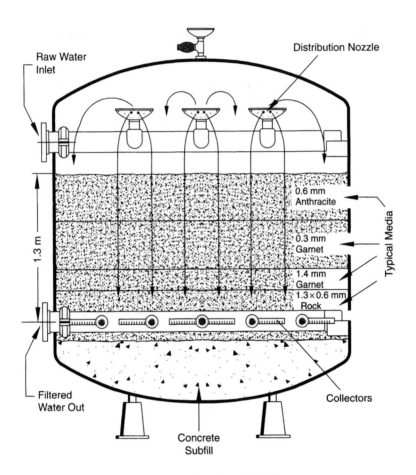

FIGURE 4.12. Deep bed down-flow (multimedia) filter.

The decision to use down-flow or up-flow filters is normally governed by the influent suspended solids content and the preferred time between backwash cycles. Down-flow filters are normally used when the suspended solids content of the influent is below 50 mg/l, and up-flow filters are used for a suspended solids content range of 50–500 mg/l. Table 4.3 provides a comparison of typical influent flow rates and solids loadings.

Granular media filters fall in the category of nonfixed-pore filters because the filter media are not held rigidly in place. Thus, if not backwashed promptly, granular media filters can unload previously filtered solids. Media migration, however, is usually not a problem because media screens are usually built into the filter vessel, preventing the media from leaving the filter vessel.

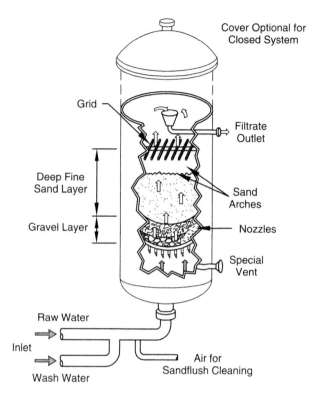

FIGURE 4.13. Deep bed up-flow filter.

TABLE 4.3
Typical parameters for granular bed filters

	Flow Rate		Solids Loading[a]	
Type	(m³/h m²)	(gpm/ft²)	(kg/m²)	(lb/ft²)
Conventional down-flow	2.4–19.6	1–8	2.4–7.3	0.5–1.5
High-rate down-flow	19.6–48.9	8–20	7.3–19.5	1.5–4
Up-flow	14.7–29.3	6–12	19.5–48.8	4–10

[a]Weight of solids trapped per unit area of media prior to backwashing.

Granular media filters use sand, gravel, anthracite, graphite, or pecan or walnut shells. The filter bed may be made of a single material or of several layers of different materials to increase the solids loading by forcing the water through progressively smaller pores.

The pore size distribution within a granular media filter is variable, depending on the random distribution of the media solids after backwashing. Because of their variable pore size, granular media filters cannot be given an absolute rating. Typically, granular media filters can consistently remove 95% of all 10 µm and larger solids.

Backwash flow rates vary with specific filter designs and are specified by the manufacturer. Some designs require an initial air or gas scour [10–15 psig (69–103 kPa) supply] to fluidize the bed. This is especially true for filters handling produced waters that contain suspended hydrocarbons that can coat the filter media. Several cycles of scour followed by flushing may be required during the backwash operation. Detergents may also be needed to aid in cleaning the filter media.

Raw water is usually used for backwash. When the backwash cycle is complete, water is allowed to flow through the filter for a period of time until the effluent quality stabilizes. Only then is the filter put back on stream.

Filters work by trapping the solid particles within their pore structure. A filter's ability to trap particles smaller than the pore space may be greatly aided by the addition of polyelectrolytes and filter aids. These chemicals promote coagulation in the line leading to the filter and aid the formation of a chemical or ionic bond between these small particles and the filter medium. For example, a specific filter may be capable of removing 90% of the 10 µm and larger particles without chemicals and 98% of the 2 µm and larger particles with chemicals.

Granular media filters are commonly used as a first filtration step (normally called "primary filtration") prior to cartridge filters (known as "secondary filtration"). This type of system works well because the granular media filter removes the bulk of the large solids, thus increasing the cycle time for replacing cartridges. The cartridge filters then remove the small solids to the required size. In addition, the cartridges catch any solids released by the sand filter due to unloading. Tables 4.4 and 4.5 provide typical operating and design parameters for two types of granular media filters.

Specific manufacturers should be contacted to select a standard granular media filter and obtain detailed sizing and operating information. To select a granular media filter, the designer should specify the following:

- maximum water flow rate,
- particle size to be removed by filtration and the percentage of removal required,
- solids concentration in the inlet water,
- design working pressure of the filter vessel, and
- maximum pressure drop available for filtration.

TABLE 4.4
Typical operating and design parameters for a specific up-flow filter

A. Operating Parameters

Service rate	14.6–29.3 m^3/h m^2 (6–12 gpm/ft^2)
Chemical treatment	Polyelectrolytes at 0.5–5 ppm
	Determine if needed by bench tests
Flush rate	Temperature-dependent (34.2–48.9 m^3/h m^2, or 14–20 gpm/ft^2)

Regeneration Time Sequence

Cycle 1:

Drain	2–5 min (drain water to top of sand bed)
Fluidize bed	5 min with air or natural gas
Flush	10–20 min (until water is clear)

Cycle 2:

Drain	3–5 min (drain water to top of sand bed)
Fluidize bed	5 min with air or natural gas
Flush	10–20 min (until water is clear)
Settle	5 min
Prefilter	15–20 min depending on water quality

B. Design Parameters

Service rate	14.6–29.3 m^3/h m^2 (6–12 gpm/ft^2) of filter area
Inlet solids	Will hold up to 49 kg of solids m^2 (10 lb/ft^2) of filter area (400 ppm maximum)
Inlet oil	Up to 50 ppm
Total outlet solids	2–5 ppm without chemical treatment
	1–2 ppm with chemical treatment
Outlet oil	Less than 1 ppm
Cycle length	2-day minimum
Fluidize gas flow	55–90 m^3/h per m^2 (3–5 cfm/ft^2) surface area (supply pressure of 83–109 kPa (12–15 psig))
Freeboard area	50–70% of total media depth
Bed expansion	Approximately 30% during flush cycle
Particle size removal	By theory, can be calculated from smallest sand (Barkman and Davidson)

C. Miscellaneous Data

1. If inlet water contains above 15 ppm oil, a solvent or surfactant wash may be required during regeneration cycle number 1.
2. Sizing of media
   1st layer: 32–38 mm gravel, 101 mm thick (1 1/4 to 1 1/2 in gravel, 4 in. thick)
   2nd layer: 10–16 mm gravel, 254 mm thick (3/8 to 5/8 in gravel, 10 in. thick)
   3rd layer: 2–3 mm sand, 305 mm thick (2–3 mm sand, 12 in. thick)
   4th layer: 1–2 mm sand, 1524 mm thick (1–2 mm sand, 60 in. thick)

TABLE 4.5
Typical operating and design parameters for a specific down-flow filter

A. Operating Parameters

Service rate	11.0 m^3/h m^2 (4.5 gpm/ft^2)
Chemical treatment	20 ppm blend of cationic polyelectrolyte and sodium laminate
Regeneration	
Backwash	4 min at 41.6 m^3/h m^2 (17 gpm/ft^2)
Rinse	4 min at 11.0 m^3/h m^2 (4.5 gpm/ft^2)

B. Design Parameters

Service rate	4.9 m^3/h m^2 (2 gpm/ft^2)
Inlet solids	<20 ppm
Inlet oil	<10 ppm

C. Miscellaneous Data

1. Sizing of media

Kind	Thickness (mm)	Thickness (in.)	Size (mm)	Size (in.)	Specific Gravity
Anthracite (top)	457	18	1.0–1.1	—	1.5
Sand	229	9	0.45–0.55	—	2.6
Garnet	76	3	0.2–0.3	—	4.2
Garnet	76	3	1.0–2.0	—	4.2
Gravel	76	3	4.8 × No. 10 Mesh	3/16 × No. 10 Mesh	2.6
Gravel	76	3	9.5 × 4.8	3/8 × 3/16	2.6
Gravel	76	3	19.0 × 9.5	3/4 × 3/8	2.6
Gravel	76	3	38.1 × 19.0	1½ × 3/4	2.6
Rock	76	3	50.8 × 38.1	2 × 1½	2.6

2. Filter may need detergent in backwash.
3. High amounts of oil during upset conditions may necessitate solvent washing the filter media.
4. Media can stick together and form balls with excessive chemicals or oil in the inlet and may require the bed to be replaced or cleaned.
5. Backwash rates in excess of 41.6 m^3/h m^2 (17 gpm/ft^2) may cause carryover of anthracite, especially when backwash water is cold.

4.7.12 Diatomaceous Earth Filters

For filtration of 0.5–1.0 μm particles, diatomaceous earth (DE) filters may be used. In the past DE filters were commonly used for removing very fine solids because they were the least costly filters available in this range. Recently, manufacturers have developed cartridge filters that can effectively remove 0.25-μm solids; this type of filter is becoming more popular than DE filters.

DE filters remove solids by forcing the water through a filter cake of diatomaceous earth. The filter cake is built up on thin wire screens of corrosion-resistant materials such as stainless steel, Monel, or inconel. A large number of wire screens, called "leaves," may be arranged within the vessel to provide a large surface area for filtration. Typically, the flow rate through DE filter screens ranges from 0.5 to 1 gpm/ft^2 (1.2–2.4 m^3/h m^2). A DE leaf filter is shown in Figures 4.14A and 4.14B. This process involves precoating the leaves with a thin layer of DE, which is introduced as slurry (see Figures 4.15A and 14.5B).

After the precoat, the water is introduced and filtration begins. A filter aid such as DE and cellulose fiber must be mixed with the water to promote an even build-up of filter cake and to maintain the filter cake's permeability. This combination is called "body feed." The weight of body feed should be roughly equal to the weight of the solids to be filtered. When the pressure drop reaches the high limit, usually between 25 and 35 psig (170 and 240 kPa), the filter cake must be backwashed from the leaves and the process started over with the precoat.

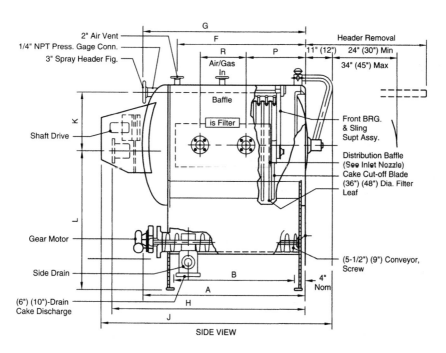

FIGURE 4.14. (A) DE filter. (Courtsey of U.S. Filter Corp.).

(continued)

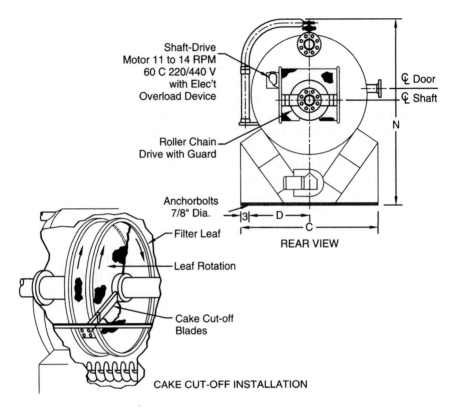

FIGURE 4.14. (B) DE filter. (Courtesy of U.S. Filter Corp.).

DE filters require slurry mixing tanks, injection pumps, and large quantities of body feed in addition to the filter vessel itself. Therefore, these systems are expensive to install and to operate, and they require much more space than do other filters.

If the precoat layer is not applied evenly to all leaves, significant amounts of solids may be released downstream. DE filters, like sand filters, are the nonfixed-pore type and suffer from unloading and media migration. In fact, unloading is typically more common with DE filters than with other filters. In addition to unloading and media migration, pressure fluctuations may cause portions of the filter cake to be lost from the leaves. The loss of filter cake then allows solids to pass downstream until the cake is again built up. Normally, guard filters are provided downstream of DE filters to protect against such leakage. Table 4.6 provides operating and design parameters for a typical DE filter.

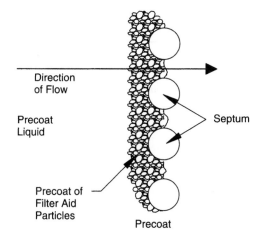

FIGURE 4.15. (A) Principles of DE filtration. (Courtesy of Johns Manville Corp.)

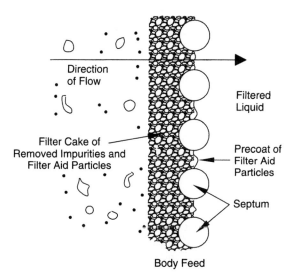

FIGURE 4.15. (B) Principles of DE filtration. (Courtesy of Johns Manville Corp.).

4.7.13 Chemical Scavenging Equipment

Chemical scavenging systems typically require chemical storage facilities, mix tanks, and injection pumps. Depending on the injection rate, the storage facilities may simply be drum racks or a small atmospheric tank. Premixed chemical scavengers may also be purchased in drums or bulk tanks if the quantities used are relatively small.

TABLE 4.6
Operating and design parameters for a typical DE filter

A. Operating Parameters

Service rate	1.2–2.4 m^3/h m^2 (0.5–1 gpm/ft^2)
DE bodyfeed	2–5 ppm DE/ppm suspended solids

Regeneration Time Sequence

Drain	1–5 min
Sluice	5 min
Fill and add precoat	3 min
Circulate	5–15 min

DE precoat:

Amount	0.5–1.0 kg/m^2 (10–20 lb/100 ft^2)
Filter slurry	30–60% water
Circulate rate	2.4–4.9 m^3/h m^2 (1–2 gpm/ft ^{2}) (4.5 fps)

B. Design Parameters

Service rate	1.2 m^3/h m^2 (0.5 gpm/ft^2)
Inlet solids	<20 ppm
Inlet oil	<10 ppm
Total outlet solids	<1 ppm

Regenerate at 20-psig pressure drop across filter

C. Miscellaneous Data

Wet bulk density DE	240–320 kg DE/m^3 (15–20 lb DE/ft^3)
Dry bulk density DE	112–240 kg DE/m^3 (7–15 lb DE/ft^3)
Specific gravity DE	2.3
Cycle length	2–3 days
Screen material	Polypropylene, plain weave, 33 × 42 count 630 deniar warp, twist direction 3.5 Z, weight 201 g/m^2 (5.92 oz/yd^2), heat set for permeability of 730 m^3/h m^2 (40 cfm/ft^2) and scoured. Stainless steel can be used.

Note: Bulk density of Perlite filter aid is one half that of DE. When Perlite is used, the above guidelines should be adjusted on an equivalent volume.

To select the best method of storage and mixing and to size injection pumps, it is necessary to calculate the chemical injection rate. The following method provides an estimate of chemical usage based on the reaction stoichiometry. Specific chemical suppliers and equipment manufacturers should be contacted to assist in making final equipment selections.

The required injection rate of chemical scavenger may be calculated as follows:

Field units

$$W_{cs} = 1.09 \times 10^{-5} Q_w SG_w C_{O_2} RMW_{cs}, \qquad (4.19a)$$

SI units

$$W_{cs} = 7.5 \times 10^{-4} Q_w SG_w C_{O_2} RMW_{cs}, \qquad (4.19b)$$

where

W_{cs} = mass flow rate of chemical scavenger, lb/day (kg/day),
Q_W = water flow rate, BWPD (m^3/h),
SG_W = water specific gravity,
C_{O_2} = inlet oxygen concentration in water, ppm,
R = stoichiometric reaction ratio between the scavenger and oxygen lb mol/hr O_2 (kg mol/h O_2),
MW_{cs} = chemical scavenger molecular weight, lb/mol.

Equations (4.19a) and (4.19b) indicate the mass flow rate of the chemical scavenger's active ingredient. The pump injection rate depends on the concentration of active ingredient in the mixed chemical solution. Chemical manufacturers can assist in determining the best solution concentration and the resulting volumetric injection rate.

The required injection rate of catalyst may be calculated as follows:

Field units

$$W_c = 7.7 \times 10^{-7} Q_w SG_w C_{ca}, \qquad (4.20a)$$

SI units

$$W_c = 5.3 \times 10^{-2} Q_w SG_w C_{ca}, \qquad (4.20b)$$

where

W_c = mass flow rate of catalyst $(CoCl_2)$, lb/day (kg/day),
C_{ca} = catalyst concentration, ppm (normally $C_{ca} = 0.001$ ppm).

Again, the volumetric injection rate depends on the mixed solution concentration of the catalyst. Manufacturers may be able to provide a premixed solution of scavenger and catalyst. This solution should be considered because it will decrease the amount of storage, mixing, and injection equipment required.

APPENDIX A

Definition of Key Water Treating Terms

A.1 Introduction

This section discusses the key terms typically used in produced water treating systems. Most of these terms are defined within Chapters 3 and 4 when initially introduced. This appendix is not intended to be a comprehensive listing of all terms.

A.2 Produced Water

The well stream from the reservoir typically contains varying quantities of water that is commonly referred to as "produced water." The produced water source can be from: (1) an aquifer layer underlying the oil and or natural gas zones, (2) connate water found within the reservoir formation sand matrix, (3) water vapor condensing from the gas phase as the result of Joule–Thompson expansion/cooling effects occurring from pressure reduction up the well bore and across wellhead chokes, (4) water-bearing formations not directly in communication with the hydrocarbon reservoir, or (5) a combination of the same. Produced water is typically salty and contains varying quantities of dissolved inorganic compounds and salts, suspended scales and other particles, dissolved gases, dissolved and dispersed liquid hydrocarbons, various organic compounds, bacteria, toxicants, and trace quantities of naturally occurring radioactive materials. Other miscellaneous sources of water from within the processing facilities (e.g., from drains, glycol regeneration units, etc.) are sometimes mixed with produced water for treatment and disposal.

A.3 Regulatory Definitions

The terminology for "total oil and grease," "dispersed oil," and "dissolved oil" may vary with location and specific test standard used by the authorities having jurisdiction. These terms should be applied with caution and should conform to the regulations and test standards applicable to the specific location.

A.4 Oil Removal Efficiency

Produced water treating equipment performance is commonly described in terms of its "oil removal efficiency." This efficiency considers only the removal of dispersed oil and neglects the dissolved oil content. For example, if the equipment removes half of the dispersed oil contained in the influent produced water, it is said to have a 50% oil removal efficiency. For a specific piece of equipment or an overall system, the oil removal efficiency can be calculated using the following equation:

$$E = 1 - \frac{C_o}{C_i} \times 100,$$

where

E = oil removal efficiency, %,
C_o = dispersed oil concentration in the water outlet (effluent) stream, ppm (mg/l),
C_i = dispersed oil concentration in the water inlet (influent) stream, ppm (mg/l).

The performance can be described by determining the inlet and outlet oil concentrations and the associated oil droplet size distributions at the equipment inlet and outlet. This information can then be used to define the oil removal efficiency for any given oil droplet size or range of droplet sizes. This concept is further discussed in Chapter 3.

A.5 Total Oil and Grease

"Total oil and grease" is defined as the combination of both the dispersed and dissolved liquid hydrocarbons and other organic compounds (i.e., "dissolved oil" plus "dispersed oil") contained in produced water. This term is referenced in certain regulatory standards and is commonly used to evaluate water treating system design. Total oil and grease consists of normal paraffinic, asphaltic, and aromatic hydrocarbon compounds plus specialty compounds from treating chemicals. The measurement of total oil and grease is dependent on the analysis method used.

A.6 Dispersed Oil

Produced water contains hydrocarbons in the form of dispersed oil droplets, which, under proper conditions, can be coalesced into a continuous hydrocarbon liquid phase and then separated from the aqueous phase using various separation devices. The diameters of these oil droplets can range from over 200 μm to less than 0.5 μm and may be surrounded by a film (emulsifier) that impedes coalescence. The relative distribution of droplet sizes is an important design parameter and is influenced by the hydrocarbon properties, temperature, downhole operating conditions, presence of trace chemical contaminants, upstream processing and pipe fittings, control valves, pumps, and other equipment that act to create turbulence and shearing action. These oil droplets are collectively defined as "dispersed oil."

Conventional water treating systems commonly used by the oil and gas industry remove only the "dispersed oil." This text focuses on the design of water treating systems that remove and recover "dispersed oil."

A.7 Dissolved Oil

Produced water contains hydrocarbons and other organic compounds that have dissolved within the aqueous phase and cannot be recovered by conventional water treating systems. Fatty acids are likely to be present within paraffinic oils and naphthenic acids within asphaltic oils. These organic acids, aromatic components, polar compounds (also called non-hydrocarbon organics), and certain treating chemicals are slightly soluble in water and collectively make up the organic compounds found in solution of the aqueous phase. The portion of these components that are dissolved into the produced water is defined as "dissolved oil." Dissolved oil is microscopically indistinguishable within the aqueous phase since it is solution at the molecular level and cannot be separated from the produced water by means of coalescence and/or gravity separation devices.

Treatment methods for removal of dissolved oil are not covered in this text. However, the oil and gas industry is currently evaluating treatment methods such as bio-treatment, air stripping, adsorption filtration, and membranes. These designs are typically prototypical in nature and require a larger capital investment, a greater maintenance work effort, and more space and may result in by-products having disposal problems more onerous than those associated with the disposal of dissolved oil. The confined space typically available on an offshore platform presents a real challenge in developing a suitable water treating process for removing dissolved oil from produced water.

A.8 Dissolved Solids

Several inorganic compounds are soluble in water. The total measure of these compounds found in solution with produced water is referred to as "total dissolved solids" (TDS). When these compounds are found in solution with the produced water, they are referred to as "dissolved solids." The most common water-soluble compound in produced water is sodium chloride. A number of other compounds collectively comprise the dissolved solids contained in produced water. These are discussed in Chapter 3.

A.9 Suspended Solids

Produced water and oil contain very small particulate solid matter held in suspension in the liquid phase by surface tension and electrostatic forces. This solid matter is referred to as a "suspended solid" and may consist of small particles of sand, clay, precipitated salts and flakes of scale, and products of corrosion such as iron oxide and iron carbonate. When suspended solids are measured by weight or volume, the composite measurement is referred to as the "total suspended solids" (TSS) content. Chapter 3 provides a detailed discussion on suspended solids.

A.10 Scale

Under certain conditions, the dissolved solids precipitate or crystallize from the produced water to form solid deposits in pipe and equipment. These solid deposits are referred to as "scale." The most common scales include calcium carbonate, calcium sulphate, barium sulphate, strontium sulphate, and iron sulfide. Scale is discussed further in Chapter 3.

A.11 Emulsion

An "emulsion" is an oil and water mixture that has been subjected to shearing resulting in the division of oil and water phases into small droplets. Most emulsions encountered in the oil field are water droplets in an oil continuous phase and are referred to as "normal emulsions." Oil droplets in a water continuous phase are referred to as "reverse emulsions." Emulsions are discussed further in Chapter 2.

APPENDIX B

Water Sampling Techniques

B.1 Sampling Considerations

Any water analysis method is only as good as the "sample" used to represent the effluent stream. Sampling of a continuously flowing stream containing two or more phases (e.g., oil and water) is difficult unless the mixture is completely emulsified or is a very fine stable dispersion. Since the sampling techniques for oil concentration measurement and particle size distribution differ in some aspects, they are described separately here.

B.2 Sample Gathering for Oil Concentration Measurement

Generally, the larger the sample the more likely it is to be representative. However, for practical reasons, the sample size varies from 15 ml to about 1 l. Typically, the smaller samples are used for daily analysis, whereas the larger samples are used for monthly regulatory compliance purposes. The smaller the residual oil droplets, the more evenly dispersed they are likely to be. Care should be exercised to avoid sampling the surface of a liquid (since this is not truly representative). "Isokinetic" (which means equal linear velocity) sampling in midstream is the best, but is rarely possible. The sample probe must be inserted so that the velocity profile remains undisturbed, thereby getting a realistic particle distribution and, thus, a higher accuracy. The general guidelines are:

- Flush the sample line thoroughly and take the sample quickly.
- Sample after a pump or a similar turbulent area where the stream is well mixed.
- Obtain the sample from a liquid-full vertical pipe, if possible.

Sample bottles should be scrupulously clean and preferably of glass. Oil or other organic material can adhere to the walls of a plastic container and give erroneous readings. Never use a metal container or a metal cap. The water can corrode it and become contaminated with corrosion products. Bottles used at oily sites or handled by an operator with oily hands can have thin surface films, and washing can leave detergent residue, both of which can give rise to erroneous and high oil readings.

General guidelines one should follow to improve measurement accuracy are as follows:

- Use only glass or inert plastic (e.g., Teflon) stoppers. Cork or other absorbent materials must not be used unless covered with aluminum foil.
- Do not rinse or overflow the bottle with the sample because an oil film will appear on the bottle and give a false reading.
- Cap the sample and prepare a label immediately with an indelible, smear-proof marking pen. Attach it to the bottle immediately.
- Analyze the entire sample and wash the bottle with solvent.

The person taking the sample must be well trained and experienced and be able to recognize a spoiled or unrepresentative sample. Samples must be correctly labeled immediately after being taken and any abnormal circumstances must be noted on the sample. If any doubt exists, the sample should be discarded and a new one taken in a fresh container.

The sampling frequency depends on the practicality of sampling at each site or may also be specified by the authorities having jurisdiction. A manned installation would require a higher-analysis frequency than an unmanned site, which may be less accessible. For example, in the United States, the EPA states, "The sample type shall be a 24-h composite consisting of the arithmetic average of results of 4 grab samples taken over a 24-h period. If only one sample is taken for any one month, it must meet both the daily and monthly limits. Samples shall be collected prior to the addition of any seawater to the produced water waste stream."

B.3 Sample Storage for Oil Concentration Measurement

If possible, perform the sample analysis as soon as the sample is obtained. If immediate analysis is not possible, as with certain samples (normally for regulatory reporting) that are sent onshore for analysis, then acidify the sample to pH 2 using hydrochloric acid (HCl) to preserve the sample against bacterial action and/or dissolve the precipitated calcium carbonate, which could cause difficulties separating the solvent phase from the water. HCl is used to avoid problems with difficulties separating the solvent phase from the water and with calcium sulfate. Acidification causes a higher total oil and grease concentration. This is because acid reacts with the organic salts to liberate organic acids, which are then extracted into the solvent. This gives a higher reading in the analysis. Therefore, when a sample has been acidified, the solvent extract should pass through a silica-gel column or similar material to remove these polar substances (organic acids).

In the case of analysis being done immediately (for daily measurement), the sample should be acidified only if the approved analysis procedure requires it.

B.4 Sample Gathering for Particle Size Analysis

The following points are applicable when obtaining a sample for particle size analysis:

- Flush the sample line thoroughly and take the sample quickly.
- Obtain the sample from a liquid-full vertical pipe, if possible.
- Use only glass or inert plastic (e.g., Teflon) stoppers. Cork or other absorbent materials must not be used unless covered with aluminum foil.
- Do not rinse or overflow the bottle with the sample as this can put an oil film on the bottle and give a false reading.
- Cap the sample and prepare a label immediately with an indelible, smear-proof marking pen. Attach it to the bottle immediately.

In addition to the above factors, other important factors that need to be considered when measuring particle size distribution are

1. *Avoid shearing of oil droplets across a sample valve.* Typically, while sampling produced water, the sample valve is never fully open, since full flow is so intense that sampling may be almost impossible. Consequently, the flow is restricted by controlling the valve partially open. In this situation, the produced water is subjected to choking from high pressure down to atmospheric conditions. The shearing within the sampling valve causes the oil droplets in the sample to break up into smaller droplets.
2. *Avoid a "dissolved gas flotation effect."* The produced water sample is depressurized as it passes through the choke valve

and gas is liberated as minute gas bubbles. These bubbles may coalesce with the oil droplets such that the oil droplets adhere to the gas bubbles and rise to the sample surface. This phenomenon is called "dissolved gas flotation." It will not alter the total concentration of oil in the sample, but could split the oil present into two separate fractions:

- Dispersed oil—remaining in the water and not affected by the gas bubbles.
- Free oil—formed as a thin film on the water surface caused by the flotation.

The free oil formed may easily adhere to the walls of the sample container as a thin and almost invisible film. This film is easily lost as the initial sample is split into subsamples. The fraction of oil in the sample ending up as free oil may be as high as 50–80%.

One way of avoiding droplet shear and gas flotation effects is to conduct an online sample measurement. This technique, however, requires specialized equipment, such as the Melvern Mastersizer, and is constrained in terms of the maximum pressure it can tolerate. Alternatively, a sample pressure cylinder (bomb) can be used to avoid droplet shear and the gas flotation effect. Here, the sample is taken in a stainless steel cylinder with a needle valve and a ball valve and thus minimizes the shearing of droplets during sampling.

3. *Avoid coalescence of oil droplets by stabilizing the sample.* A sample used for droplet size measurement may need to be stabilized to avoid coalescence of the small oil droplets to larger ones. The propensity of a droplet to coalesce increases as the oil concentration of the sample increases. This stabilization becomes more important when measuring samples with a high oil concentration. Stabilization may be achieved by diluting the sample with a known amount of water. This reduces the chance of droplet coalescence and also stabilizes the sample by reducing the salt concentration of the sample. Further stabilization of the sample may be achieved by the addition of a viscous polymer solution and/or surfactant (e.g., 2% sodium dodecyl sulphate). However, one should be very cautious while adding such chemicals since the wrong surfactant could actually promote coalescence.

4. *Sample may contain solids and other non-particles in addition to the oil droplets.* Produced water samples often contain solids and other non-oil particles in addition to oil droplets. To determine only the oil droplet size distribution, one must first determine the size distribution for all particles within

the sample and then determine the size distribution of particles left behind in the sample after solvent extraction. Solvent extraction removes all of the oil droplets from the sample, leaving behind only the solids and non-extractable non-oil particles. One can then block out those size ranges (corresponding to the solids and non-oil particles) from the initial size distribution to obtain a more representative size distribution.

APPENDIX C

Oil Concentration Analysis Techniques

C.1 Introduction

Several analytical techniques measure the amount of oil and grease in water. These techniques may be broadly classified as either gravimetric or infrared (IR) absorbance methods and are described in detail here. These methods are based on the extraction of oil and grease into a solvent. A sample may contain suspended solids, which have to be filtered. In this case, the sample must first undergo solvent extraction followed by filtration of the extract.

Several different solvents have been used. These include petroleum ether, diethyl ether, chloroform, and carbon tetrachloride. Of these, petroleum ether and diethyl ether are highly flammable, whereas chloroform (although a very good solvent) and carbon tetrachloride are toxic. Thus, these solvents are not recommended for use. Currently, 1,1,2 trichloro, trifluoroethane (Freon 113) is used when infrared (IR) absorbance is used for analysis. However, these solvents are being phased out because of potential interference with the ozone layer in the atmosphere. Studies are currently under way to find a replacement solvent. Potential candidates include hexane, cyclohexane, methylene chloride, perchloroethylene, and a commercial hydro chlorofluorocarbon (DuPont 123). When the gravimetric technique is used for analysis, 1,1,1-trichloroethane or dichloroethylene may also be used.

"Total oil and grease" is defined by the measurement procedure stipulated by the authorities having jurisdiction. Important variations that could give different results for the same sample are:

- *Number of extractions performed on the water sample.* Multiple extractions with Freon on the same water sample will generally give a higher oil concentration than a single extraction.
- *The solvent-to-sample ratio.* A higher solvent-to-sample ratio will also give a higher oil concentration.
- *Determination of IR absorbance at multiple wavelengths.* This variation will give a higher oil concentration as opposed to absorbance measurement at a single wavelength.
- *Use of silica gel.* This variation is discussed below.

C.2 Determination of Dissolved Oil and Grease

The dissolved oil and grease content is first determined by measuring the total oil and grease content and then subtracting the measured dispersed oil and grease content. The measured dispersed oil is obtained by removal of the dissolved oil and grease from the solvent with silica gel. This can be expressed by the following equation:

$$\text{dissolved oil and grease} = \text{total oil and grease} - \text{dispersed oil and grease.} \qquad \text{(C.1)}$$

Dissolved organic matter is generally either polar or of low molecular weight. To remove the dissolved organic matter, the solvent extract is contacted (either in an adsorption column or by intimate mixing) with activated silica gel (e.g., Florisil) or alumina. The materials not adsorbed by the silica gel are described as "dispersed" oil and grease.

In the absence of silica gel, filtration can be used to remove the dispersed oil and grease. In this technique the filtrate, obtained after passing through a 0.45-μm filter paper is left behind on the filter paper. The U.S. EPA uses the term "petroleum hydrocarbons" for dispersed oil and grease.

The choice of measuring total oil and grease versus dispersed oil and grease is important from an operational standpoint because conventional water treating technology can reduce the concentration of dispersed oil and grease, but not the concentration of dissolved oil and grease. However, measurement techniques are strictly specified in some countries (e.g., the United States) and left open to negotiation or operator discretion in others.

C.3 Gravimetric Method

This method is the U.S. EPA-required method to measure oil and grease in produced water for regulatory compliance purposes in the United States. A detailed procedure can be found in EPA 413.1.

In this method, the water sample is extracted with Freon 113 and the extract evaporated to remove the solvent. The weight of the residue is related to the concentration of oil in the water sample. Generally, for the same water sample, this method gives a lower value for oil and grease concentration than the IR absorbance method because of the loss of volatile organics during the evaporation process.

C.3.1 Advantages

- Stipulated by the EPA as the technique to be used to measure oil in produced water discharge for regulatory purposes.
- Method is simple and well understood.

C.3.2 Disadvantages

- Requires samples to be collected and preserved according to EPA protocol for shipment to an onshore laboratory.
- Is time-consuming.
- Uses Freon (a CFC) as a solvent.
- Lower limit of measurement is 5 mg/l.
- Not applicable to light hydrocarbons that volatilize below 70 °C.

C.4 Infrared (IR) Absorbance Method

For infrared absorbance methods (e.g., EPA 413.2 and EPA 418.1), the water sample is extracted with Freon 113. The IR absorbance of the extract is measured at single or multiple wavelength(s) to give the oil concentration. In this method, the water sample is often acidified to prevent any salts from precipitating out (e.g., iron sulfide). IR absorbance at multiple wavelengths results in a higher oil concentration measurement than if a single wavelength was used.

C.4.1 Advantages

- Fast and convenient for offshore surveillance.
- The lower limit of measurement is 0.2 mg/l.

C.4.2 Disadvantages

- Uses Freon (a CFC).
- The lower limit is 0.2 mg/l.

C.5 Analysis of Variance of Analytical Results

In 1975, the U.S. EPA had a set of six different samples analyzed by a number of laboratories using both the gravimetric and IR methods. Scatter in reported analysis included sampling errors as well as analytical errors. The true values were taken to be the average of the reported values (excluding those of extreme scatter) and are as follows:

Sample	1	2	3	4	5	6
Gravimetric	201.7	209.3	95.5	77.6	33.7	19.1
Infrared	261.3	265.4	131.0	110.0	50.7	23.0

(*Note*: Error is defined as the difference of an observation from the best obtainable estimate of the true value, which in this case is the arithmetic mean.)

It is interesting to note that the gravimetric value is lower than the infrared value of each of the samples. This is expected since the solvent evaporation step in the gravimetric process causes some loss of volatile organics, leading to lower results than the IR method. Errors ranged from 0–241%. The worst laboratory had errors between 49% and 98%, whereas the best laboratory had errors up to 8%.

C.6 Particle Size Analysis

The oil droplet size distribution is one of the key parameters influencing water treating equipment selection. Therefore, accurate measurement of the oil droplet size distribution is an important task. Another important parameter is quantifying the size distribution upstream and downstream of production equipment, such as control valves.

Oil and other particles in produced water range in size from less than 1 µm up to several hundred microns. Although many particles found in produced water are not spherical, for practical purposes, the particles are represented by equivalent spheres.

C.7 Droplet Size Measurement Equipment

Three different types of equipment are commonly used for droplet size measurement. Each has its advantages and disadvantages. First, establish the information desired before selecting the equipment type:

1. *Coulter Counter.* The Coulter Counter consists of two electrodes immersed in a beaker of sample water, which contains a sufficient number of dissolved ions to easily conduct an electrical current. The negative electrode is located inside a glass tube, which is sealed except for a tiny hole or orifice on the side of the tube. The positive electrode is located in the water sample beaker. A constant electrical current is

passed from the positive electrode to the negative electrode through the orifice. When a non-conductive particle passes through the orifice, a change in electrical resistance occurs between the two electrodes which is proportional to the particle volume.

A fixed volume of water containing suspended particles is forced through the orifice. As each particle passes through the orifice, the increased resistance results in a voltage that is proportional to the particle volume. The series of pulses produced by a series of particles passing through the orifice are electronically scaled and counted, yielding a particle size distribution. One must realize that the particle "diameter" given by the counter is the diameter of a fictitious equivalent sphere with the same volume as the real particle.

This equipment has some limitations because the size range is limited. In addition, the samples have to be suspended in an electrolyte solution, which can prove difficult if the sample is totally soluble in the solution. However, the Coulter Counter does provide both frequency and volume distributions against volumetric particle size.

2. *Light (laser) scattering counters.* These include instruments that are based on the principle of light absorption/total scatter, or light blockage, to detect particles in a fluid. Water flows through a sensor cell, and as each particle passes through the intense beam of light in the sensor, light is scattered. The instrument measures the magnitude of each scattered light pulse, which is proportional to the surface area of the particle. The particle diameter determined by the instrument in this case is the diameter of a sphere with the same surface area as the particle.

   Laser diffraction systems are widely accepted due to their ease of use, wide size range, and simple sample preparation. However, as an optical technique, it is still subject to variations in response from particle shape and refractive index and is unable to give frequency information (that is, the number of particles within a given size range). However, these techniques can provide relative frequency information (that is, percent of total particle volume within a given size range).

3. *Microscopy.* In this technique, the droplet size distribution is determined by observing the water sample under a microscope and visually measuring the size of the droplets. Often a magnified photograph is also used for visual determination. This technique has the advantage of being able to distinguish between oil droplets and non-oil particles. A microscope also helps to see first hand if there are any extreme shape factors.

However, the technique generally uses a very small sample volume and therefore may not be representative.

References

Kawahara, F. K., "A study to Select a Suitable Replacement Solvent for Freon 113 in the Gravimetric Determination of Oil and Grease," EPA, 2 October 1991.

Patton, C. C., "Water Sampling and Analysis," *Applied Water Technology*, Campbell Petroleum Series, Norman, OK 1986.

"Particle Sizing—Past and Present," *Particle Sizing Review*, Filtration and Separation Journal, July/August 1993.

API RP 45, *"Recommended Practice for Analysis of Oil-Field Waters,"* API, Washington, DC 1981.

Index

Printed and bound by CPI Group (UK) Ltd, Croydon, CR0 4YY

08/05/2025

01864823-0001